Physical Science

DATASHEETS FOR LABBOOK

HOLT, RINEHART AND WINSTON

A Harcourt Classroom Education Company

Austin • New York • Orlando • Atlanta • San Francisco • Boston • Dallas • Toronto • London

To the Student

Does it take more force to lift a book or to slide the book up an inclined plane? How does the density of a liquid affect whether the liquid sinks or floats? You can explore these questions and many more in the *Holt Science & Technology Physical Science* LabBook. These activities present practical applications of the science concepts you are learning in class.

Datasheets for LabBook is a reproduction of the LabBook found at the end of your *Holt Science & Technology Physical Science* textbook, with one difference—the datasheets provide plenty of space for you to record your data, observations, and answers. This design allows you to keep all the relevant information in one place as you complete each laboratory exercise. The datasheets introduce you to the scientific method and walk you through each activity and experiment step by step. Charts, tables, and graphs are included to make data collection and analysis easier, and space is provided for you to write your observations and conclusions.

Art and Photo Credits
All work, unless otherwise noted, contributed by Holt, Rinehart and Winston.
Abbreviated as follows: (t) top; (b) bottom; (l) left; (r) right; (c) center; (bkgd) background.
Front cover (owl), Kim Taylor/Bruce Coleman, Inc.; (bridge), Henry K. Kaiser/Leo de Wys; (dove), Stephen Dalton/Photo Researchers, Inc.; Page 9 (cr), Romark Illustrations; 63 (l), Preface, Inc.; 102 (c), Stephen Durke/Washington Artists

Printed in the United States of America

ISBN 0-03-054404-1   4 5 6 **862** 04 03 02

▪ CONTENTS ▪

CONTENTS, CONTINUED

CONTENTS, CONTINUED

Name _____ Date _____ Class_____

1 **STUDENT WORKSHEET**

DISCOVERY LAB

Exploring the Unseen

Your teacher will give you a box in which a special divider has been created. Your task is to describe this divider as precisely as possible—without opening the box! Your only aid is a marble that is also inside the box. This task will allow you to demonstrate your understanding of the scientific method. Good luck!

MATERIALS
• a sealed mystery box

SCIENTIFIC **METHOD**

Ask a Question

1. In the space below, record the question that you are trying to answer by doing this experiment. (Hint: Read the introductory paragraph again if you are not sure what your task is.)

_____

_____

Form a Hypothesis

2. Before you begin the experiment, think about what's required. Do you think you will be able to easily determine the shape of the divider? What about its texture? Its color? In the space below, write a hypothesis that states how much you think you will be able to determine about the divider during the experiment. (Remember, you can't open the box!)

_____

_____

_____

Test the Hypothesis

3. Using all the methods you can think of (except opening the box), test your hypothesis. Make careful notes about your tests and observations in your ScienceLog.

Analyze the Results

4. What characteristics of the divider were you able to identify?

_____

_____

In the margin areas of this page, draw or write your best description of the interior of the box.

Exploring the Unseen, continued

5. Do your observations support your hypothesis? Explain. If your results do not support your hypothesis, write a new hypothesis and test it.

_____

_____

_____

_____

_____

6. With your teacher's permission, open the box and look inside. Record your observations below.

_____

_____

_____

_____

_____

Communicate Results

7. Write a paragraph summarizing your experiment. Be sure to include your methods, whether your results supported your hypothesis, and how you could improve your methods.

_____

_____

_____

_____

_____

_____

_____

_____

_____

Name _____ Date _____ Class_____

2 **STUDENT WORKSHEET**

Off to the Races!

Scientists often use models—representations of objects or systems. Physical models, such as a model airplane, are generally a different size than the objects they represent. In this lab you will build a model car, test its design, and then try to improve the design.

MATERIALS

- 2 sheets of typing paper
- glue
- 16 cm clothes-hanger wire
- pliers or wire cutters
- metric ruler
- rubber eraser or wooden block
- ramp (board and textbooks)
- stopwatch

Procedure

1. Using the materials listed, design and build a car that will carry the load (the eraser or block of wood) down the ramp as quickly as possible. Your car must be no wider than 8 cm, it must have room to carry the load, and it must roll.

2. As you test your design, do not be afraid to rebuild or redesign your car. Improving your methods is an important part of scientific progress.

3. When you have a design that works well, measure the time required for your car to roll down the ramp. Record this time in the space provided next to "Round 1" in the data table below. Repeat this step several times.

Timed Trials

	Trial 1 (sec)	Trial 2 (sec)	Trial 3 (sec)	Trial 4 (sec)
Round 1				
Round 2				

4. Try to improve your model. Find one thing that you can change to make your model car roll faster down the ramp. Write a description of the change in the space below.

_____

_____

_____

_____

_____

_____

5. Repeat step 3, entering your data for Round 2 in the spaces provided in the data table above.

_____

Off to the Races! continued

_____

Analysis

6. Why is it important to have room in the model car for the eraser or wood block? (Hint: Think about the function of a real car.)

_____

_____

7. Before you built the model car, you created a design for it. Do you think this design is also a model? Explain.

_____

_____

_____

8. Based on your observations in this lab, list three reasons why it is helpful for automobile designers to build and test small model cars rather than immediately building a full-size car.

_____

_____

_____

_____

_____

9. In this lab, you built a model that was smaller than the object it represented. Some models are larger than the objects they represent. List three examples of larger models that are used to represent objects. Why is it helpful to use a larger model in these cases?

_____

_____

_____

_____

_____

_____

_____

Name _____ Date _____ Class _____

3 **STUDENT WORKSHEET**

Measuring Liquid Volume

In this lab you will use a graduated cylinder to measure and transfer precise amounts of liquids. Remember, in order to accurately measure liquids in a graduated cylinder, you should read the level at the bottom of the meniscus, the curved surface of the liquid.

MATERIALS

- masking tape
- marker
- 6 large test tubes
- test-tube rack
- 10 mL graduated cylinder
- 3 beakers filled with colored liquids
- small funnel

Procedure

1. Using the masking tape and marker, label the test tubes A, B, C, D, E, and F. Place them in the test-tube rack. Be careful not to confuse the test tubes.

2. Using the 10 mL graduated cylinder and the funnel, pour 14 mL of the red liquid into test tube A. (To do this, first pour 10 mL of the liquid into the test tube and then add 4 mL of liquid.)

3. Rinse the graduated cylinder and funnel between uses.

4. Measure 13 mL of the yellow liquid, and pour it into test tube C. Then measure 13 mL of the blue liquid, and pour it into test tube E.

5. Transfer 4 mL of liquid from test tube C into test tube D. Transfer 7 mL of liquid from test tube E into test tube D.

6. Measure 4 mL of blue liquid from the beaker, and pour it into test tube F. Measure 7 mL of red liquid from the beaker, and pour it into test tube F.

7. Transfer 8 mL of liquid from test tube A into test tube B. Transfer 3 mL of liquid from test tube C into test tube B.

Liquid Data

Collect Data

8. In the data table at left, record the color of the liquid in each test tube.

9. Use the graduated cylinder to measure the volume of liquid in each test tube, and record the volumes in your data table.

10. Record your color observations in a table of class data prepared by your teacher. Make a data table in the space available on the next page, and record all the color observations for your class.

Test tube	Color	Volume (mL)
A		
B		
C		
D		
E		
F		

Analyze the Results

11. Did all of the groups report the same colors? Explain why the colors were the same or different.

_____

_____

_____

_____

_____

_____

12. Why should you not fill the graduated cylinder to the very top?

_____

_____

_____

_____

_____

_____

_____

Name _____ Date _____ Class_____

Coin Operated

All pennies are exactly the same, right? Probably not! After all, each penny was made in a certain year at a specific mint, and each has traveled a unique path to reach your classroom. But all pennies *are* similar. In this lab you will investigate differences and similarities among a group of pennies.

MATERIALS

- 10 pennies
- metric balance
- a few sheets of paper
- 100 mL graduated cylinder
- water
- paper towels

Procedure

1. Place one penny next to each of the numbers on the table. To make the pennies fit, you may need to alternate their placement above and below the table.

2. Use the metric balance to find the mass of each penny to the nearest 0.1 g. Record the mass of each penny in Data Table 1 below.

3. On a table your teacher will provide, make a mark in the correct column of the table for each penny you measured.

Data Table 1

Penny	1	2	3	4	5	6	7	8	9	10
Mass (g)										

4. Separate your pennies into two piles based on the class data. Place each pile on its own sheet of paper.

5. Measure and record the mass of each pile. Write the masses in Data Table 2 below.

6. Fill a graduated cylinder about halfway with water. Carefully measure the volume and record it in the table.

7. Carefully place the pennies from one pile in the graduated cylinder. Measure and record the new volume in the table.

Data Table 2

Pile	Volume of water (mL)	Volume of water + pennies (mL)	Volume of pennies (cm^3)	Density (g/cm^3)
1				
2				

Coin Operated, continued

8. Carefully remove the pennies from the graduated cylinder, and dry them off.

9. Repeat steps 6 through 8 for each pile of pennies.

Analyze the Results

10. Determine the volume of the displaced water by subtracting the initial volume from the final volume. This amount is equal to the volume of the pennies. Record the volume of the pennies in the table.

11. Calculate the density of the pile. To do this, divide the total mass of the pennies by the volume of the pennies. Record the density in the appropriate space in the table.

Draw Conclusions

12. How is it possible for the pennies to have different densities?

_____

_____

_____

_____

_____

13. What clues might allow you to separate the pennies into the same groups without experimentation? Explain.

_____

_____

_____

_____

_____

Name _____ Date _____ Class_____

5 STUDENT WORKSHEET

Volumania!

You have learned how to measure the volume of a solid object that has square or rectangular sides. But there are lots of objects in the world that have irregular shapes. In this lab activity, you'll learn some ways to find the volume of objects that have irregular shapes.

MATERIALS
Part A • graduated cylinder • water • various small objects supplied by your teacher **Part B** • bottom half of a 2 L plastic bottle or similar container • water • aluminum pie pan • paper towels • funnel • graduated cylinder

Part A: Finding the Volume of Small Objects

Procedure

1. Fill a graduated cylinder half full with water. Read the volume of the water, and record it in the table below. Be sure to look at the surface of the water at eye level and to read the volume at the bottom of the meniscus, as shown at right.

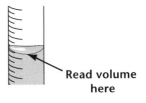

Read volume here

2. Carefully slide one of the objects into the tilted graduated cylinder.

3. Read the new volume, and record it in the data table.

4. Subtract the old volume from the new volume. The resulting amount is equal to the volume of the solid object. Record the volume of the solid object in the data table.

5. Use the same method to find the volume of the other objects. Record your results in the data table.

Volume Calculation Chart

Name of object	Starting volume (mL)	Ending volume (mL)	Volume of object (cm³)

Volumania! continued

Analysis

6. What changes do you have to make to the volumes you determine in order to express them correctly?

_____

_____

_____

_____

7. Do the heaviest objects always have the largest volumes? Why or why not?

_____

_____

_____

_____

Part B: Finding the Volume of Your Hand

Procedure

Refer to the figure at the top of page 631 in your textbook.

8. Completely fill the container with water. Put the container in the center of the pie pan. Be sure not to spill any of the water into the pie pan.

9. Make a fist and put your right hand into the container up to your wrist.

10. Remove your hand and let the excess water drip into the container, not the pie pan. Dry your hand with a paper towel.

11. Use the funnel to pour the overflow water into the graduated cylinder. Measure the volume. This is the volume of your hand. Record the volume in the appropriate space below. (Remember to use the correct unit of volume for a solid object.)

12. Repeat this procedure with your left hand.

Volume of right hand	
Volume of left hand	

_____

Volumania! continued

_____

Analysis

13. Was the volume the same for both of your hands? If not, were you surprised?

_____

_____

_____

What might account for a person's hands having different volumes?

_____

_____

_____

14. Would it have made a difference if you had placed your open hand into the container instead of your fist? Explain your reasoning.

_____

_____

_____

15. Compare the volume of your right hand with the volume of your classmates' right hands. In your ScienceLog, create a class graph of right-hand volumes. What is the average right-hand volume for your class?

_____

Going Further

- Design an experiment to determine the volume of a person's body. In your plans, be sure to include the materials needed for the experiment and the procedures that must be followed. Include a sketch that shows how your materials and methods would be used in this experiment.
- Using an encyclopedia, the Internet, or other reference materials, find out how the volumes of very large samples of matter—such as an entire planet—are determined.

CHAPTER 2

6 **STUDENT WORKSHEET**

Determining Density

The density of an object is its mass divided by its volume. But how does the density of a small amount of a substance relate to the density of a larger amount of the same substance? In this lab, you will calculate the density of one marble and of a group of marbles. Then you will confirm the relationship between the mass and volume of a substance.

Collect Data

1. Use the table below to record your data.

Data Table

Mass of marble (g)	Total mass of marbles (g)	Total volume (mL)	Volume of marbles (cm³)	Density of marbles (g/cm³)

MATERIALS

- 100 mL graduated cylinder
- water
- paper towels
- 8 to 10 glass marbles
- metric balance
- graph paper

2. Fill the graduated cylinder with 50.0 mL of water. If you put in too much water, twist one of the paper towels and use its end to absorb excess water.

3. Measure the mass of a marble as accurately as you can (to at least one-tenth of a gram). Record the marble's mass in the first column of the table. Write the same value in the second column.

4. Carefully drop the marble into the tilted cylinder, and measure the total volume. Record the volume in the third column. Fill in the rest of the row.

5. Measure and record the mass of another marble. Add the masses of the marbles together, and record this value in the second column of the table.

Determining Density, continued

6. Carefully drop the marble into the graduated cylinder without removing the previous marble. Complete the row of information in the table.

7. Repeat steps 5 and 6, adding one marble at a time. Stop when you run out of marbles, the water no longer completely covers the marbles, or the graduated cylinder is full.

Analyze the Results

8. Examine the data in your table. As the number of marbles increases, what happens to the total mass of the marbles? What happens to the volume of the marbles? What happens to the density of the marbles?

_____

_____

_____

_____

_____

_____

9. Graph the total mass of the marbles (*y*-axis) versus the volume of the marbles (*x*-axis) in the grid on the next page.

Is the graph a straight line or a curved line?

_____

_____

Draw Conclusions

10. Does the density of a substance depend on the amount of substance present? Explain how your results support your answer.

_____

_____

_____

_____

_____

▲ CHAPTER 2

Determining Density, continued

Graph of Mass Versus Volume

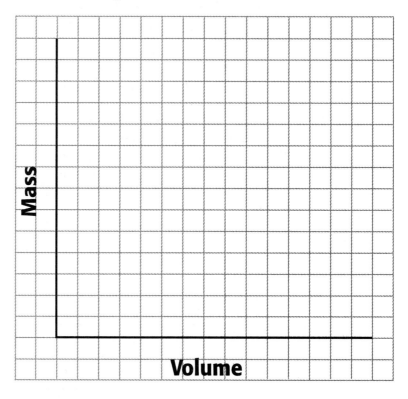

Going Further

Calculate the slope of the graph. How does the slope compare with values in the column titled "Density of marbles"? Explain.

_____

_____

_____

_____

_____

_____

_____

_____

DATASHEET

7 **STUDENT WORKSHEET**

DISCOVERY LAB

Layering Liquids

You have learned that liquids form layers according to their densities. In this lab, you'll discover whether it matters in which order you add the liquids.

MATERIALS
• liquid A
• liquid B
• liquid C
• beaker or other small, clear container
• 10 mL graduated cylinders (3)
• 3 funnels

SCIENTIFIC METHOD

Make a Prediction

1. Does the order in which you add liquids of different densities to a container affect the order of the layers formed by those liquids?

_____

Conduct an Experiment

2. Using the graduated cylinders, add 10 mL of each liquid to the clear container. Remember to read the volume at the bottom of the meniscus. Record the order in which you added the liquids.

_____

_____

3. Observe the liquids in the container. Sketch what you see in the space to the left. Be sure to label the layers and the colors.

4. Add 10 mL more of liquid C. Observe what happens, and write your observations below.

_____

_____

5. Add 20 mL more of liquid A. Observe what happens, and write your observations below.

_____

_____

Analyze Your Results

6. Which of the liquids has the greatest density?

_____

_____

Which has the least density?

_____

_____

CHAPTER 2

Layering Liquids, continued

How can you tell?

_____

_____

_____

7. Did the layers change position when you added more of liquid C? Explain your answer.

_____

_____

_____

8. Did the layers change position when you added more of liquid A? Explain your answer.

_____

_____

_____

Communicate Your Results

9. Find out in what order your classmates added the liquids to the container. Compare your results with those of a classmate who added the liquids in a different order. Were your results different? In the space below, explain why or why not.

_____

_____

_____

Draw Conclusions

10. Based on your results, evaluate your prediction from step 1.

_____

_____

_____

_____

_____

DATASHEET

8 **STUDENT WORKSHEET**

SKILL BUILDER

White Before Your Eyes

You have learned how to describe matter based on its physical and chemical properties. You have also learned some clues that can help you determine whether a change in matter is a physical change or a chemical change. In this lab, you'll use what you have learned to describe four substances based on their properties and the changes they undergo.

MATERIALS

- 4 spatulas
- baking powder
- plastic-foam egg carton
- 3 eyedroppers
- water
- stirring rod
- vinegar
- iodine solution
- baking soda
- cornstarch
- sugar

CHAPTER 2 ▲ ▲ ▲

Procedure

1. Use Table 1 on the next page to record your observations during steps 2 through 6 of this lab.

2. Use a spatula to place a small amount (just enough to cover the bottom of the cup) of baking powder into three cups of your egg carton. Look closely at the baking powder, and record observations of its color, texture, etc., in the column of Table 1 titled "Unmixed."

3. Use an eyedropper to add 60 drops of water to the baking powder in the first cup. Stir with the stirring rod. Record your observations in Table 1 in the column titled "Mixed with water." Clean your stirring rod.

4. Use a clean dropper to add 20 drops of vinegar to the second cup of baking powder. Stir. Record your observations in Table 1 in the column titled "Mixed with vinegar." Clean your stirring rod.

5. Use a clean dropper to add five drops of iodine solution to the third cup of baking powder. Stir. Record your observations in the column in Table 1 titled "Mixed with iodine solution." Clean your stirring rod. **Caution:** Be careful when using iodine. Iodine will stain your skin and clothes.

6. Repeat steps 2–5 for each of the other substances. Use a clean spatula for each substance.

Analysis

7. In Table 2, write the type of change you observed and state the property that the change demonstrates.

8. What clues did you use to identify when a chemical change happened?

_____

_____

_____

_____

White Before Your Eyes, continued

_____

Table 1: Observations

Substance	Unmixed	Mixed with water	Mixed with vinegar	Mixed with iodine solution
Baking powder				
Baking soda				
Cornstarch				
Sugar				

White Before Your Eyes, continued

Table 2: Changes and Properties

Substance	Mixed with water		Mixed with vinegar		Mixed with iodine solution	
	Change	Property	Change	Property	Change	Property
Baking powder	chemical	reactivity with water				
Baking soda						
Cornstarch						
Sugar						

CHAPTER 2

DATASHEET

9 STUDENT WORKSHEET

Full of Hot Air!

Why do hot-air balloons float gracefully above Earth, while balloons you blow up fall to the ground? The answer has to do with the density of the air inside the balloon. Density is mass per unit volume, and volume is affected by changes in temperature. In this experiment, you will investigate the relationship between the temperature of a gas and its volume. Then you will be able to determine how the temperature of a gas affects its density.

MATERIALS

- 2 aluminum pans
- water
- metric ruler
- hot plate
- ice water
- balloon
- 250 mL beaker
- heat-resistant gloves

SCIENTIFIC METHOD

Form a Hypothesis

1. How does an increase or decrease in temperature affect the volume of a balloon?

_____

_____

Test the Hypothesis

2. Fill an aluminum pan with water about 4 to 5 cm deep. Put the pan on the hot plate, and turn the hot plate on.

3. While the water is heating, fill the other pan 4 to 5 cm deep with ice water.

4. Blow up a balloon inside the 250 mL beaker, as shown in the figure on page 636 of your textbook. The balloon should fill the beaker but should not extend outside the beaker. Tie the balloon at its opening.

5. Place the beaker and balloon in the ice water. Observe what happens. Record your observations.

_____

_____

6. Remove the balloon and beaker from the ice water. Observe the balloon for several minutes. Record any changes.

_____

_____

7. Put on heat-resistant gloves. When the hot water begins to boil, put the beaker and balloon in the hot water. Observe the balloon for several minutes and record your observations.

_____

_____

Full of Hot Air! continued

8. Turn off the hot plate. When the water has cooled, carefully pour it into a sink.

Analyze the Results

9. Summarize your observations of the balloon. Relate your observations to Charles's law.

_____

_____

_____

_____

10. Was your hypothesis for step 1 supported? If not, revise your hypothesis.

_____

_____

_____

Draw Conclusions

11. Based on your observations, how is the density of a gas affected by an increase or decrease in temperature?

_____

_____

_____

_____

12. Explain in terms of density and Charles's law why heating the air allows a hot-air balloon to float.

_____

_____

_____

_____

_____

CHAPTER 3

Name _____ Date _____ Class _____

SKILL BUILDER

Can Crusher

Condensation can occur when gas particles come near the surface of a liquid. The gas particles slow down because they are attracted to the liquid. This reduction in speed causes the gas particles to condense into a liquid. In this lab, you'll see that particles that have condensed into a liquid don't take up as much space and therefore don't exert as much pressure as they did in the gaseous state.

MATERIALS
• water
• 2 empty aluminum cans
• heat-resistant gloves
• hot plate
• tongs
• 1 L beaker

Conduct an Experiment

1. Place just enough water in an aluminum can to slightly cover the bottom.

2. Put on heat-resistant gloves. Place the aluminum can on a hot plate turned to the highest temperature setting.

3. Heat the can until the water is boiling. Steam should be rising vigorously from the top of the can.

4. Using tongs, quickly pick up the can and place the top 2 cm of the can upside down in the 1 L beaker filled with room-temperature water.

5. Describe your observations below.

_____

_____

_____

Analyze the Results

6. The can was crushed because the atmospheric pressure outside the can became greater than the pressure inside the can. Explain what happened inside the can to cause this.

_____

_____

_____

_____

_____

_____

_____

_____

Can Crusher, continued

Draw Conclusions

7. Inside every popcorn kernel is a small amount of water. When you make popcorn, the water inside the kernels is heated until it becomes steam. Explain how the popping of the kernels is the opposite of what you saw in this lab. Be sure to address the effects of pressure in your explanation.

_____

_____

_____

_____

_____

_____

_____

_____

_____

Going Further

Try the experiment again, but use ice water instead of room-temperature water. Explain your results in terms of the effects of temperature.

_____

_____

_____

_____

_____

_____

_____

_____

_____

_____

▲▲ CHAPTER 3
▲

DATASHEET

11 **STUDENT WORKSHEET**

DISCOVERY LAB

A Hot and Cool Lab

When you add energy to a substance through heating, does the substance's temperature always go up? When you remove energy from a substance through cooling, does the substance's temperature always go down? In this lab you'll investigate these important questions with a very common substance—water.

MATERIALS

Part A
- 250 or 400 mL beaker
- water
- heat-resistant gloves
- hot plate
- thermometer
- stopwatch
- graph paper

Part B
- 100 mL graduated cylinder
- water
- large coffee can
- crushed ice
- rock salt
- thermometer
- wire-loop stirring device
- stopwatch
- graph paper

Part A: Boiling Water

Make a Prediction

1. What happens to the temperature of boiling water when you continue to add energy through heating?

_____

_____

Procedure

2. Fill the beaker about one-third to one-half full with water.

3. Put on heat-resistant gloves. Turn on the hot plate, and put the beaker on the burner. Put the thermometer in the beaker. **Caution:** Be careful not to touch the burner.

Collect Data

4. In the table below, record the temperature of the water every 30 seconds. Continue doing this until about one-fourth of the water boils away. Note the first temperature reading at which the water is steadily boiling.

Table 1: Temperature Over Time

Time (s)	30	60	90	120	150	180	210	240	270	300	330
Temp (°C)											
Time (s)	360	390	420	450	480	510	540	570	600	630	660
Temp (°C)											

5. Turn off the hot plate.

6. While the beaker is cooling, make a graph of temperature (*y*-axis) versus time (*x*-axis). Draw an arrow pointing to the first temperature at which the water was steadily boiling.

A Hot and Cool Lab, continued

7. After you finish the graph, use heat-resistant gloves to pick up the beaker. Pour the warm water out, and rinse the warm beaker with cool water. **Caution:** Even after cooling, the beaker is still too warm to handle without gloves.

Part B: Freezing Water
Make Another Prediction

8. What happens to the temperature of freezing water when you continue to remove energy through cooling?

_____

_____

Procedure

9. Put approximately 20 mL of water in the graduated cylinder.

10. Put the graduated cylinder in the coffee can and fill in around the graduated cylinder with crushed ice. Pour rock salt on the ice around the graduated cylinder. Place the thermometer and the wire-loop stirring device in the graduated cylinder.

11. As the ice melts and mixes with the rock salt, the level of ice will decrease. Add ice and rock salt to the can as needed.

Collect Data

12. In the table below, record the temperature of the water in the graduated cylinder every 30 seconds. Stir the water with the stirring device. **Caution:** Do not stir with the thermometer.

Table 2: Temperature Over Time

Time (s)	30	60	90	120	150	180	210	240	270	300	330
Temp (°C)											
Time (s)	360	390	420	450	480	510	540	570	600	630	660
Temp (°C)											

13. Once the water begins to freeze, stop stirring. Do not try to pull the thermometer out of the solid ice in the cylinder.

CHAPTER 3

A Hot and Cool Lab, continued

14. Note the temperature when you first notice ice crystals forming in the water. Continue taking readings until the water in the graduated cylinder is completely frozen.

15. Make a graph of temperature (*y*-axis) versus time (*x*-axis). Draw an arrow to the temperature reading at which the first ice crystals form in the water in the graduated cylinder.

Analyze the Results (Parts A and B)

16. What does the slope of each graph represent?

_____

_____

17. How does the slope when the water is boiling compare with the slope before the water starts to boil?

_____

_____

_____

Why is the slope different for the two periods?

_____

_____

_____

18. How does the slope when the water is freezing compare with the slope before the water starts to freeze?

_____

_____

_____

A Hot and Cool Lab, continued

Why is the slope different for the two periods?

_____

_____

_____

_____

_____

Draw Conclusions (Parts A and B)

19. Addition or subtraction of energy leads to changes in the movement of particles that make up solids, liquids, and gases. Use this idea to explain why the temperature graphs of the two experiments look the way they do.

_____

_____

_____

_____

_____

_____

_____

_____

_____

_____

_____

_____

_____

_____

▲ CHAPTER 3

DATASHEET

12 STUDENT WORKSHEET

DISCOVERY LAB

Flame Tests

Fireworks produce fantastic combinations of color when they are ignited. The different colors are the results of burning different compounds. Imagine that you are the lead chemist for a fireworks company. The label has fallen off one of the boxes that is filled with a compound, and you must identify the unknown compound so that it may be used in the correct fireworks display. To identify the compound, you will use your knowledge that every compound has a unique set of properties.

MATERIALS

- 4 small test tubes
- test-tube rack
- masking tape
- 4 chloride test solutions
- spark igniter
- Bunsen burner
- wire and holder
- dilute hydrochloric acid in a small beaker
- distilled water in a small beaker

SCIENTIFIC **METHOD**

Make a Prediction

1. Can you identify the unknown compound by heating it in a flame? Explain.

_____

_____

Conduct an Experiment

Caution: Be very careful in handling all chemicals. Tell your teacher immediately if you spill a chemical.

2. Arrange the test tubes in the test-tube rack. Use masking tape to label the tubes with the following names: calcium chloride, potassium chloride, sodium chloride, and unknown.

3. Ask your teacher for your portions of the solutions.

4. Light the burner. Clean the wire by dipping it into the dilute hydrochloric acid and then into distilled water. Holding the wooden handle, heat the wire in the blue flame of the burner until the wire is glowing and it no longer colors the flame. **Caution:** Use extreme care around an open flame.

Collect Data

5. Dip the clean wire into the first test solution. Hold the wire at the tip of the inner cone of the burner flame. In the table below, record the color given to the flame.

Test Results

Compound	Color of flame
Calcium chloride	
Potassium chloride	
Sodium chloride	
Unknown	

Flame Tests, continued

6. Clean the wire by repeating step 4.

7. Repeat steps 5 and 6 for the other solutions.

8. Follow your teacher's instructions for cleanup and disposal.

Analyze the Results

9. Is the flame color a test for the metal or for the chloride in each compound? Explain your answer.

_____

_____

_____

_____

_____

_____

_____

_____

10. What is the identity of your unknown solution? How do you know?

_____

_____

_____

_____

_____

_____

Draw Conclusions

11. Why is it necessary to carefully clean the wire before testing each solution?

_____

_____

_____

_____

CHAPTER 4

▲ ▲
▲
▲

Flame Tests, continued

12. Would you expect the compound sodium fluoride to produce the same color as sodium chloride in a flame test?
Why or why not?

_____

_____

_____

_____

_____

_____

_____

13. Each of the compounds you tested is made from chlorine,
which is a poisonous gas at room temperature. Why is it
safe to use these compounds without a gas mask?

_____

_____

_____

_____

_____

_____

DATASHEET

13   STUDENT WORKSHEET

DISCOVERY
LAB

A Sugar Cube Race!

If you drop a sugar cube into a glass of water, how long will it take to dissolve? Will it take 5 minutes, 10 minutes, or longer? What can you do to speed up the rate at which it dissolves? Should you change something about the water, the sugar cube, or the process? In other words, what variable should you change? Before reading further, make a list of variables that could be changed in this situation.

_____

_____

MATERIALS

- water
- graduated cylinder
- 2 sugar cubes
- 2 beakers or other clear containers
- clock or stopwatch
- other materials approved by your teacher

SCIENTIFIC
METHOD

Make a Prediction

1. Choose one variable to test. Record your choice below, and predict how changing your variable will affect the rate of dissolving.

_____

_____

_____

Conduct an Experiment

2. Pour 150 mL of water into one of the beakers. Add one sugar cube, and use the stopwatch to measure how long it takes for the sugar cube to dissolve. You must not disturb the sugar cube in any way! Record this time in the table.

Data Table

Dissolve time–undisturbed	Dissolve time–variable

3. Tell your teacher how you wish to test the variable. Do not proceed without his or her approval. You may need additional equipment.

4. Prepare your materials to test the variable you have picked. When you are ready, start your procedure for speeding up the dissolving of the sugar cube. Use the stopwatch to measure the time. Record this time in the table.

Analyze the Results

5. Compare your results with the results from step 2. Was your prediction correct? Why or why not?

_____

_____

_____

▲▲ CHAPTER 4 ▲

A Sugar Cube Race! continued

Draw Conclusions

6. Why was it necessary to observe the sugar cube dissolving on its own before you tested the variable?

_____

_____

_____

_____

_____

7. Do you think that changing more than one variable would speed up the rate of dissolving even more? Explain your reasoning.

_____

_____

_____

_____

_____

_____

_____

Communicate Results

8. Discuss your results with a group that tested a different variable. Which variable had a greater effect on the rate of dissolving?

_____

_____

_____

_____

_____

DATASHEET

14 **STUDENT WORKSHEET**

Making Butter

A colloid is an interesting substance. It has properties of both solutions and suspensions. Colloidal particles are not heavy enough to settle out, so they remain evenly dispersed throughout the mixture. In this activity, you will make butter—a very familiar colloid—and observe the characteristics that classify butter as a colloid.

MATERIALS
• a marble
• small, clear container with lid
• heavy cream
• clock or stopwatch

Procedure

1. Place a marble inside the container, and fill the container with heavy cream. Put the lid tightly on the container.

2. Take turns shaking the container vigorously and constantly for 10 minutes. Record the time when you begin shaking the container. Every minute, stop shaking the container and hold it up to the light. Record your observations.

_____

_____

_____

_____

_____

_____

3. Continue shaking the container, taking turns if necessary. When you see, hear, or feel any changes inside the container, note the time and change.

_____

_____

_____

4. After 10 minutes of shaking, you should have a lump of "butter" surrounded by liquid inside the container. Describe both the butter and the liquid in detail.

_____

_____

_____

_____

_____

CHAPTER 4

_____

Making Butter, continued

_____

5. Let the container sit for about 10 minutes. Observe the butter and liquid again, and record your observations.

_____

_____

_____

_____

Analysis

6. When you noticed the change in the container, what did you think was happening at that point?

_____

_____

_____

_____

_____

7. Based on your observations, explain why butter is classified as a colloid.

_____

_____

_____

_____

_____

8. What kind of mixture is the liquid that is left behind? Explain.

_____

_____

_____

_____

_____

DATASHEET

15 **STUDENT WORKSHEET**

Unpolluting Water

In many cities, the water supply comes from a river, lake, or reservoir. This water may include several mixtures, including suspensions (with suspended dirt, oil, or living organisms) and solutions (with dissolved chemicals). To make the water safe to drink, your city's water supplier must remove impurities. In this lab, you will model the procedures used in real water-treatment plants.

MATERIALS

- "polluted" water
- graduated cylinder
- 250 mL beakers (4)
- 2 plastic spoons
- small nail
- 8 oz plastic-foam cups (2)
- scissors
- 2 pieces of filter paper
- washed fine sand
- metric ruler
- washed activated charcoal
- rubber band

Part A: Untreated Water

Procedure

1. Measure 100 mL of "polluted" water into a graduated cylinder. Be sure to shake the bottle of water before you pour so your sample will include all the impurities.

2. Pour the contents of the graduated cylinder into a beaker.

3. Record your observations of the water in the "Before treatment" column of the table on the next page.

Part B: Settling In

If a suspension is left standing, the suspended particles will settle to the top or bottom. You should see a layer of oil at the top.

Procedure

4. Separate the oil by carefully pouring the oil into another beaker. You can use a plastic spoon to get the last bit of oil from the water. Record your observations in the table at the top of page 36.

Part C: Filtration

Cloudy water can be a sign of small particles still in suspension. These particles can usually be removed by filtering. Water-treatment plants use sand and gravel as filters.

Procedure

5. Make a filter as follows:
   a. Use the nail to poke 5 to 10 small holes in the bottom of one of the cups.
   b. Cut a circle of filter paper to fit inside the bottom of the cup. (This will keep the sand in the cup.)
   c. Fill the cup to 2 cm below the rim with wet sand. Pack the sand tightly.
   d. Set the cup inside an empty beaker.

CHAPTER 4

Unpolluting Water, continued

Observations

	Before treatment	After oil separation	After sand filtration	After charcoal
Color				
Clearness				
Odor				
Any layers?				
Any solids?				
Water volume?				

6. Pour the polluted water on top of the sand, and let it filter through. Do not pour any of the settled mud onto the sand. (Dispose of the mud as instructed by your teacher.) In the table, record your observations of the water collected in the beaker.

Part D: Separating Solutions

Something that has been dissolved in a solvent cannot be separated using filters. Water-treatment plants use activated charcoal to absorb many dissolved chemicals.

Procedure

7. Place activated charcoal about 3 cm deep in the unused cup. Pour the water collected from the sand filtration into the cup, and stir for a minute with a spoon.

8. Place a piece of filter paper over the top of the cup, and fasten it in place with a rubber band. With the paper securely in place, pour the water through the filter paper and back into a clean beaker. Record your observations in the table.

Analysis (Parts A–D)

9. Is your unpolluted water safe to drink? Why or why not?

_____

_____

_____

_____

_____

_____

_____

_____

_____

_____

_____

_____

10. When you treat a sample of water, do you get out exactly the same amount of water that you put in? Explain.

_____

_____

_____

_____

11. Some groups may still have cloudy water when they finish. Explain a possible cause for this.

_____

_____

_____

_____

_____

CHAPTER 4

DATASHEET

16 **STUDENT WORKSHEET**

Built for Speed

DESIGN
YOUR OWN

Imagine that you are an engineer at GoCarCo, a toy-vehicle company. GoCarCo is trying to beat the competition by building a new toy vehicle. Several new designs are being tested. Your boss has given you one of the new toy vehicles and instructed you to measure its speed as accurately as possible with the tools you have. Other engineers (your classmates) are testing the other designs. Your results could decide the fate of the company!

MATERIALS
• toy vehicle
• meterstick
• masking tape
• stopwatch

Procedure

1. How will you accomplish your goal? Write a paragraph to describe your goal and your procedure for this experiment. Be sure that your procedure includes several trials.

_____

_____

_____

_____

_____

_____

2. Show your plan to your boss (teacher). Get his or her approval to carry out your procedure.

3. Perform your stated procedure. Create a data table in the space below and record all of your data. Be sure to express all data in the correct units.

Built for Speed, continued

Analysis

4. What was the average speed of your vehicle? How does your result compare with the results of the other engineers?

_____

_____

_____

5. Compare your technique for determining the speed of your vehicle with the techniques of the other engineers. Which technique do you think is the most effective?

_____

_____

_____

_____

6. Was your toy vehicle the fastest? Explain why or why not.

_____

_____

_____

_____

_____

_____

Going Further

Think of several conditions that could affect your vehicle's speed. Design an experiment to test your vehicle under one of those conditions. Write a paragraph to explain your procedure. Be sure to include an explanation of how that condition changes your vehicle's speed.

_____

_____

_____

_____

▲ **CHAPTER 5**

DATASHEET

Detecting Acceleration

Have you ever noticed that you can "feel" acceleration? In a car or in an elevator you notice the change in speed or direction—even with your eyes closed! Inside your ears are tiny hair cells. These cells can detect the movement of fluid in your inner ear. When you accelerate, the fluid does, too. The hair cells detect this acceleration in the fluid and send a message to your brain. This allows you to sense acceleration.

In this activity you will build a device that detects acceleration. Even though this device is made with simple materials, it is very sensitive. It registers acceleration only briefly. You will have to be very observant when using this device.

MATERIALS

- scissors
- string
- 1 L container with watertight lid
- pushpin
- small cork or plastic-foam ball
- modeling clay
- water

Procedure

1. Cut a piece of string that is just long enough to reach three quarters of the way inside the container.

2. Use a pushpin to attach one end of the string to the cork or plastic-foam ball.

3. Use modeling clay to attach the other end of the string to the center of the *inside* of the container lid. Be careful not to use too much string—the cork (or ball) should hang no farther than three-quarters of the way into the container.

4. Fill the container to the top with water.

5. Put the lid tightly on the container with the string and cork (or ball) on the inside.

6. Turn the container upside down (lid on the bottom). The cork should float about three-quarters of the way up inside the container. You are now ready to use your accelerometer to detect acceleration.

7. Put the accelerometer lid side down on a tabletop. Notice that the cork floats straight up in the water.

8. Now gently start pushing the accelerometer across the table at a constant speed. Notice that the cork quickly moves in the direction you are pushing then swings backward. If you did not see this happen, try the same thing again until you are sure you can see the first movement of the cork.

9. Once you are familiar with how to use your accelerometer, try the following changes in motion, and record your observations of the cork's first motion for each change.

    a. While moving the device across the table, push a little faster.

    _____

    _____

_____

Detecting Acceleration, continued

b. While moving the device across the table, slow down.

_____

_____

c. While moving the device across the table, change the direction that you are pushing. (Try changing to both the left and to the right.)

_____

_____

d. Make any other changes in motion you can think of. You should only change one part of the motion at a time.

_____

_____

Analysis

10. The cork moves forward (in the direction you were pushing the bottle) when you speed up but backward when you slow down. Why? (Hint: Think about the direction of acceleration.)

_____

_____

_____

_____

_____

_____

_____

_____

_____

Detecting Acceleration, continued

11. When you push the bottle at a constant speed, why does the cork quickly swing back after it shows the direction of acceleration?

_____

_____

_____

_____

12. Imagine you are standing on a corner, watching a car that is waiting at a stoplight. A passenger inside the car is holding some helium balloons. Based on what you observed with your accelerometer, what do you think will happen to the balloons when the car begins moving?

_____

_____

_____

_____

_____

_____

_____

_____

Going Further

If you move the bottle in a circle at a constant speed, what do you predict the cork will do? Try it, and check your answer.

_____

_____

_____

_____

_____

_____

Name _____ Date _____ Class_____

18 **STUDENT WORKSHEET**

Science Friction

In this experiment, you will investigate three types of friction—static, sliding, and rolling—to determine which is the largest force and which is the smallest force.

MATERIALS

- scissors
- string
- textbook (covered)
- spring scale (force meter)
- 3 to 4 wooden or metal rods

Ask a Question

1. Which type of friction is the largest force—static, sliding, or rolling? Which is the smallest?

Form a Hypothesis

2. Write a statement that answers the questions above. Explain your reasoning.

_____

_____

_____

Test the Hypothesis/Collect Data

3. Cut a piece of string and tie it in a loop that fits in your textbook, as shown on page 650 of your textbook. Hook the string to the spring scale.

4. Practice the next three steps several times before you collect data.

5. To measure the static friction between the book and the table, pull the spring scale very slowly. Record the largest force on the scale before the book starts to move.

_____

_____

6. After the book begins to move, you can determine the sliding friction. Record the force required to keep the book sliding at a slow, constant speed.

_____

_____

7. Place two or three rods under the book to act as rollers. Make sure the rollers are evenly spaced. Place another roller in front of the book so that the book will roll onto it. Pull the force meter slowly. Measure the force needed to keep the book rolling at a constant speed.

_____

_____

Analyze the Results

8. Which type of friction was the largest?

_____

_____

Which was the smallest?

_____

_____

9. Do the results support your hypothesis? If not, how would you revise or retest your hypothesis?

_____

_____

_____

_____

_____

Communicate Results

10. Compare your results with those of another group. Are there any differences?

_____

_____

_____

Working together, design a way to improve the experiment and resolve possible differences.

_____

_____

_____

_____

_____

_____

Name _____ Date _____ Class_____

Relating Mass and Weight

Why do objects with more mass weigh more than objects with less mass? All objects have weight on Earth because their mass is attracted by Earth's gravitational force. Because the mass of an object on Earth is constant, the relationship between the mass of an object and its weight is also constant. You will measure the mass and weight of several objects to verify the relationship between mass and weight on the surface of Earth.

MATERIALS

- metric balance
- small classroom objects
- spring scale (force meter)
- string
- scissors
- graph paper

Collect Data

1. Record your data in the table below.

2. Using the metric balance, find the mass of five or six small classroom objects designated by your teacher. Record the masses in the table below.

Mass and Weight Measurements

Object	Mass (g)	Weight (N)

3. Using the spring scale, find the weight of each object, and record the values in the table. (You may need to use the string to hang some objects from the spring scale.)

Analyze the Results

4. Using your data, construct a graph of weight (*y*-axis) versus mass (*x*-axis). Draw a line that best fits all your data points.

5. Does the graph confirm the relationship between mass and weight on Earth? Explain your answer.

_____

_____

_____

_____

CHAPTER 5

DATASHEET

20 STUDENT WORKSHEET

A Marshmallow Catapult

Catapults use projectile motion to launch objects across distances. A variety of factors can affect the distance an object can be launched, such as the weight of the object, how far the catapult is pulled back, and the catapult's strength. In this lab, you will build a simple catapult and determine the angle at which the catapult will launch an object the farthest.

MATERIALS
• plastic spoon
• block of wood, 3.5 × 3.5 × 1 cm
• duct tape
• miniature marshmallows
• protractor
• meterstick

Form a Hypothesis

1. At what angle, from 10° to 90°, will a catapult launch a marshmallow the farthest?

_____

_____

Test the Hypothesis

2. Examine the data table below.

Data Table

Angle	Distance 1 (cm)	Distance 2 (cm)	Average distance (cm)
10°			
20°			
30°			
40°			
50°			
60°			
70°			
80°			
90°			

3. Attach the plastic spoon to the 1 cm side of the block with duct tape.

4. Place one marshmallow in the center of the spoon, and tape it to the spoon. This serves as a ledge to hold the marshmallow that will be launched.

CHAPTER 6

A Marshmallow Catapult, continued

5. Line up the bottom corner of the block with the bottom center of the protractor, as shown in the photograph on page 652 of the textbook. Start with the block at 10°.

6. Place a marshmallow in the spoon, on top of the taped marshmallow. Pull back lightly and let go. Measure and record the distance from the catapult that the marshmallow lands. Repeat the measurement, and calculate an average. Record these values in the data table.

7. Repeat step 6 for each angle up to 90°.

Analyze the Results

8. At what angle did the catapult launch the marshmallow the farthest? Compare this with your hypothesis. Explain any differences.

_____

_____

_____

Draw Conclusions

9. Does the path of an object's projectile motion depend on the catapult's angle? Support your answer with your data.

_____

_____

_____

_____

10. At what angle should you throw a ball or shoot an arrow so that it will fly the farthest? Why? Support your answer with your data.

_____

_____

_____

_____

_____

DATASHEET

21 STUDENT WORKSHEET

MAKING MODELS

Blast Off!

You have been hired as a rocket scientist for NASA. Your job is to design a rocket that will have a controlled flight while carrying a payload. Keep in mind that Newton's laws will have a powerful influence on your rocket.

MATERIALS

- duct tape
- 3 m fishing line
- pencil
- small paper cup
- 15 cm length of string (2)
- long, thin balloon
- twist tie
- drinking straw
- meterstick
- 100 pennies

Procedure

1. When you begin your experiment, your teacher will tape one end of the fishing line to the ceiling.

2. Use a pencil to poke a small hole in each side of the cup near the top. Place a 15 cm piece of string through each hole, and tape down the ends inside.

3. Inflate the balloon, and use the twist tie to hold it closed.

4. Tape the free ends of the strings to the sides of the balloon near the bottom. The cup should hang below the balloon. Your model rocket should look like a hot-air balloon.

5. Thread the fishing line that is hanging from the ceiling through the straw. Tape the balloon securely to the straw.

6. Tape the loose end of the fishing line to the floor.

Collect Data

7. Untie the twist tie while holding the end of the balloon closed. When you are ready, release the end of the balloon. Mark and record the maximum height of the rocket. Create a data table in the space below.

8. Repeat the procedure, adding a penny to the cup each time until your rocket cannot lift any more pennies.

_____

Blast Off! continued

Analysis

9. In a paragraph, describe how all three of Newton's laws influenced the flight of your rocket.

_____

_____

_____

_____

_____

_____

_____

_____

_____

_____

_____

_____

10. Draw a diagram of your rocket in your ScienceLog. Label the action and reaction forces.

Going Further

Brainstorm ways to modify your rocket so that it will carry the most pennies to the maximum height. Select the best design. When your teacher has approved all the designs, each team will build and launch their rocket. Which variable did you modify? How did this variable affect the flight of the rocket?

_____

_____

_____

_____

_____

_____

_____

_____

DATASHEET

22 **STUDENT WORKSHEET**

SKILL
BUILDER

Inertia-Rama!

Inertia is a property of all matter, from small particles of dust to enormous planets and stars. In this lab, you will investigate the inertia of various shapes and types of matter. Keep in mind that each investigation requires you to either overcome or use the object's inertia.

MATERIALS
Station 1
• hard-boiled egg
• raw egg
Station 2
• quarter
• index card
• cup
Station 3
• spool of thread
• suspended mass
• scissors
• meterstick

Station 1: Magic Eggs

Procedure

1. There are two eggs at this station—one is hard-boiled (solid all the way through) and the other is raw (liquid inside). The masses of the two eggs are about the same. The eggs are marked so that you can tell them apart.

2. You will spin each egg and then stop it from spinning by placing a finger on its center. Before you do anything to either egg, write some predictions. Which egg will be the easiest to spin? Which egg will be the easiest to stop?

   _____

   _____

3. Spin the hard-boiled egg. Then place your finger on it to make it stop spinning. What did you observe?

   _____

4. Repeat step 3 with the raw egg.

   _____

5. Compare your predictions with your observations. (Repeat steps 3 and 4 if necessary.)

   _____

   _____

Analysis

6. Explain why the eggs behave differently when you spin them even though they should have the same inertia. (Hint: Think about what happens to the liquid inside the raw egg.)

   _____

   _____

   _____

Inertia-Rama! continued

7. In terms of inertia, explain why the eggs react differently when you try to stop them.

_____

_____

_____

_____

_____

_____

Station 2: Coin in a Cup

Procedure

8. At this station, you will find a coin, an index card, and a cup. Place the card over the cup. Then place the coin on the card over the center of the cup.

9. Write down a method for getting the coin into the cup without touching the coin and without lifting the card.

_____

_____

_____

_____

_____

10. Try your method. If it doesn't work, try again until you find a method that does work. When you are done, place the card and coin on the table for the next group.

Analysis

11. Use Newton's first law of motion to explain why the coin falls into the cup when your method is used.

_____

_____

_____

_____

_____

---

12. Explain why pulling on the card slowly will not work, even though the coin has inertia. (Hint: Friction is a force.)

_____

_____

_____

_____

Station 3: The Magic Thread

Procedure

13. At this station, you will find a spool of thread and a mass hanging from a strong string. Cut a piece of thread about 40 cm long. Tie the thread around the bottom of the mass.

14. Pull gently on the end of the thread. What happens? Record your observations below.

_____

_____

_____

_____

_____

_____

15. Stop the mass from moving. Now hold the end of the thread so that there is a lot of slack between your fingers and the mass.

16. Give the thread a quick, hard pull. You should observe a very different event. Record your observations below.

_____

_____

_____

_____

_____

_____

Analysis

17. Use Newton's first law of motion to explain why the results of a gentle pull are so different from the results of a hard pull.

_____

_____

_____

_____

_____

_____

Draw Conclusions

18. Remember that both moving and nonmoving objects have inertia. Explain why it is as hard to throw a bowling ball as it is to catch a thrown bowling ball.

_____

_____

_____

_____

_____

19. Why is it harder to run with a backpack full of books than with an empty backpack?

_____

_____

_____

_____

_____

_____

DATASHEET

23 STUDENT WORKSHEET

Quite a Reaction

Catapults have been used for centuries to throw objects great distances. You may already be familiar with catapults after doing the marshmallow catapult lab. According to Newton's third law of motion (whenever one object exerts a force on a second object, the second object exerts an equal and opposite force on the first), when an object is launched, something must also happen to the catapult. In this activity, you will build a kind of catapult that will allow you to observe the effects of Newton's third law of motion and the law of conservation of momentum.

MATERIALS

- glue
- 10 × 15 cm rectangles of cardboard (3)
- 3 pushpins
- string
- rubber band
- 6 plastic straws
- a marble
- scissors
- meterstick

Conduct an Experiment

1. Glue the cardboard rectangles together to make a stack of three.

2. Push two of the pushpins into the cardboard stack near the corners at one end, as shown on page 656 of your textbook. These will be the anchors for the rubber band.

3. Make a small loop of string.

4. Put the rubber band through the loop of string, and then place the rubber band over the two pushpin anchors. The rubber band should be stretched between the two anchors with the string loop in the middle.

5. Pull the string loop toward the end of the cardboard stack opposite the end with the anchors, and fasten the loop in place with the third pushpin.

6. Place the six straws about 1 cm apart on a tabletop or on the floor. Then carefully center the catapult on top of the straws.

7. Put the marble in the closed end of the *V* formed by the rubber band.

8. Use scissors to cut the string holding the rubber band, and observe what happens. (Be careful not to let the scissors touch the cardboard catapult when you cut the string.)

9. Reset the catapult with a new piece of string. Try launching the marble several times to be sure that you have observed everything that happens during a launch. Record all your observations below.

_____

_____

_____

_____

Analyze the Results

10. Which has more mass, the marble or the catapult?

_____

11. What happened to the catapult when the marble was launched?

_____

12. How far did the marble fly before it landed?

_____

_____

13. Did the catapult move as far as the marble did?

_____

_____

Draw Conclusions

14. Explain why the catapult moved backward.

_____

_____

_____

_____

15. If the forces that made the marble and the catapult move apart are equal, why didn't the marble and the catapult move apart the same distance? (Hint: The fact that the marble can roll after it lands is not the answer.)

_____

_____

_____

_____

_____

16. The momentum of an object depends on the mass and velocity of the object. What is the momentum of the marble before it is launched? What is the momentum of the catapult? Explain your answers.

_____

_____

_____

17. Using the law of conservation of momentum, explain why the marble and the catapult move in opposite directions after the launch.

_____

_____

_____

_____

_____

Going Further

How would you modify the catapult if you wanted to keep it from moving backward as far as it did? (It still has to rest on the straws.) Using items that you can find in the classroom, design a catapult that will move backward less than the original design. Draw your design below.

Name _____ Date _____ Class_____

24 STUDENT WORKSHEET

SKILL BUILDER

Fluids, Force, and Floating

Why do some objects sink in fluids while others float? In this lab, you'll get a sinking feeling as you determine that an object floats when its weight is less than the buoyant force exerted by the surrounding fluid.

MATERIALS

- large rectangular tank or plastic tub
- water
- metric ruler
- small rectangular baking pan
- labeled masses
- metric balance
- paper towels

Procedure

1. Examine the data table below.

Data Table

Measurement	Trial 1	Trial 2
Length (l), cm		
Width (w), cm		
Initial height (h_1), cm		
Initial volume (V_1), cm^3 $V_1 = l \times w \times h_1$		
New height (h_2), cm		
New volume (V_2), cm^3 $V_2 = l \times w \times h_2$		
Displaced volume (ΔV), cm^3 $\Delta V = V_2 - V_1$		
Mass of displaced water, g $m = \Delta V \times 1$ g/cm^3		
Weight of displaced water, N (buoyant force)		
Weight of pan and masses, N		

2. Fill the tank or tub half full with water.

3. Measure (in centimeters) the length, width, and initial height of the water. Record your measurements in the data table.

4. Using the equation given in the data table, determine the initial volume of water in the tank. Record your results in the data table.

5. Place the pan in the water, and place masses in the pan. Keep adding masses until the pan sinks to about three-quarters of its height. This will cause the water level in the tank to rise. Record the new height of the water. Then use this value to determine and record the new volume of water.

6. Determine the volume of the water that was displaced by the pan and masses, and record this value in the table. The displaced volume is equal to the new volume minus the initial volume.

7. Determine the mass of the displaced water by multiplying the displaced volume by its density (1 g/cm^3). Record the mass in the data table.

8. Divide the mass by 100. The value you get is the weight of the displaced water in newtons (N). This is equal to the buoyant force. Record the weight of the displaced water in the data table.

9. Remove the pan and masses, and determine their total mass (in grams) using the balance. Convert the mass to weight (N), as you did in step 8. Record the weight of the masses and pan in the data table.

10. Place the empty pan back in the tank. Perform a second trial by repeating steps 5–9. This time add masses until the pan is just about to sink.

Analysis

11. Compare the buoyant force (the weight of the displaced water) with the weight of the pan and masses for both trials.

_____

_____

_____

_____

12. How did the buoyant force differ between the two trials? Explain.

_____

_____

_____

_____

Fluids, Force, and Floating, continued

13. Based on your observations, what would happen if you were to add even more mass to the pan than you did in the second trial? Explain your answer in terms of the buoyant force.

_____

_____

_____

_____

_____

_____

14. What would happen if you put the masses in the water without the pan? What difference does the pan's shape make?

_____

_____

_____

_____

_____

CHAPTER 7

25 STUDENT WORKSHEET

Density Diver

Crew members of a submarine can control the submarine's density underwater by allowing water to flow into and out of special tanks. These changes in density affect the submarine's position in the water. In this lab, you'll control a "density diver" to learn for yourself how the density of an object affects its position in a fluid.

MATERIALS

- 2 L plastic bottle with screw-on cap
- water
- medicine dropper

SCIENTIFIC **METHOD**

Form a Hypothesis

1. How does the density of an object determine whether the object floats, sinks, or maintains its position in a fluid? Write your hypothesis below.

_____

_____

_____

Test the Hypothesis

2. Completely fill the 2 L plastic bottle with water.

3. Fill the diver (medicine dropper) approximately halfway with water, and place it in the bottle. The diver should float with only part of the rubber bulb above the surface of the water. If the diver floats too high, carefully remove it from the bottle and add a small amount of water to the diver. Place the diver back in the bottle. If you add too much water and the diver sinks, empty out the bottle and diver and go back to step 2.

4. Put the cap on the bottle tightly so that no water leaks out.

5. Apply various pressures to the bottle. Carefully watch the water level inside the diver as you squeeze and release the bottle. Record your observations below.

_____

_____

_____

6. Try to make the diver rise, sink, or stop at any level. Record your technique and your results.

_____

_____

_____

Density Diver, continued

Analyze the Results

7. How do the changes inside the diver affect its position in the surrounding fluid?

_____

_____

_____

_____

8. What is the relationship between the water level inside the diver and the diver's density? Explain.

_____

_____

_____

_____

Draw Conclusions

9. What relationship did you observe between the diver's density and the diver's position in the fluid?

_____

_____

_____

_____

10. Explain how your density diver is like a submarine.

_____

_____

_____

_____

CHAPTER 7

11. Explain how pressure on the bottle is related to the diver's density. Be sure to include Pascal's principle in your explanation.

_____

_____

_____

_____

_____

_____

_____

12. What was the variable in this experiment? What factors were controlled?

_____

_____

_____

_____

_____

_____

DATASHEET

26 STUDENT WORKSHEET

MAKING MODELS

Taking Flight

When air moves above and below the wing of an airplane, the air pressure below the wing is higher than the air pressure above the wing. This creates lift. In this activity, you will build a model airplane to help you identify how wing size and thrust (forward force provided by the engine) can affect the lift needed for flight.

MATERIALS
• sheet of paper

1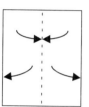

2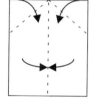

3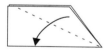

4

5

6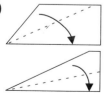

Procedure

1. Fold the paper in half lengthwise and open it again, as shown at left. Make sure to crease all folds well.

2. Fold the right- and left-hand corners toward the center crease.

3. Fold the entire sheet in half along the center crease.

4. With the plane lying on its side, fold the top front edge down so that it meets the bottom edge, as shown.

5. Fold the top wing down again, bringing the top edge to the bottom edge.

6. Turn the plane over, and repeat steps 4 and 5 for the other wing.

7. Raise both wings away from the body to a position slightly above horizontal. Your plane is ready!

8. Point the plane slightly upward, and gently throw it. Repeat several times. Describe your observations below. **Caution:** Be sure to point the plane away from people.

_____

_____

_____

9. Make the wings smaller by folding them one more time. Gently throw the plane overhand. Repeat several times. Describe your observations below.

_____

_____

_____

_____

_____

Taking Flight, continued

10. Try to achieve the same flight path you saw when the plane's wings were bigger. Record your technique.

_____

_____

_____

_____

Analysis

11. What happened to the plane's flight when you reduced the size of its wings? Explain.

_____

_____

_____

_____

_____

12. What provided your airplane's thrust?

_____

_____

_____

13. From your observations, how does changing the thrust affect the plane's lift?

_____

_____

_____

_____

_____

DATASHEET

27 **STUDENT WORKSHEET**

A Powerful Workout

Does the amount of work you do depend on how fast you do it? No! But doing work in a shorter amount of time does affect your power—the rate at which work is done. In this lab, you'll calculate your work and power when climbing a flight of stairs at different speeds. Then you'll compare your power with that of an ordinary household object—a 100 W light bulb.

MATERIALS
• flight of stairs
• metric ruler
• stopwatch

Ask a Question

1. How does your power when climbing a flight of stairs compare with the power of a 100 W light bulb?

Form a Hypothesis

2. Write a hypothesis that answers the question in step 1. Explain your reasoning.

_____

_____

3. Use the following table to record your data.

Table 1: Data Collection

Height of step (cm)	Number of steps	Height of stairs (m)	Time for slow walk (s)	Time for quick walk (s)

Test the Hypothesis

4. Measure the height of one stair step. Record the measurement in Table 1.

5. Count the number of stairs, including the top step, and record this number in Table 1.

6. Calculate the height (in meters) of the stairs by multiplying the number of steps by the height of one step. Record your answer. (You will need to convert from centimeters to meters.)

7. Using a stopwatch, measure how many seconds it takes you to walk slowly up the flight of stairs. Record your measurement in Table 1.

8. Now measure how many seconds it takes you to walk quickly up the flight of stairs. Be careful not to overexert yourself.

CHAPTER 8

Analyze the Results

9. You will record the results of your calculations in the table below.

Table 2: Work and Power Calculations

Weight (N)	Work (J)	Power for slow walk (W)	Power for quick walk (W)

10. Determine your weight in newtons by multiplying your weight in pounds (lb) by 4.45 N/lb. Record it in Table 2.

11. Calculate and record the work you've done to climb the stairs using the following equation:

$$\text{work} = \text{force} \times \text{distance}$$

Remember that 1 N•m is 1 J. (**Hint:** Remember that force is expressed in newtons.)

12. Calculate and record your power for each trial (the slow walk and the quick walk) using the following equation:

$$\text{power} = \frac{\text{work}}{\text{time}}$$

Remember that the unit for power is the watt (1 W = 1 J/s).

A Powerful Workout, continued

Draw Conclusions

13. In step 11 you calculated your work done in climbing the stairs. Why didn't you calculate your work for each trial?

_____

_____

14. Look at your hypothesis in step 2. Was your hypothesis supported? Explain.

_____

_____

Write a statement that describes how your power in each trial compares with the power of a 100 W light bulb.

_____

_____

_____

_____

_____

Where is work done in a light bulb?
Electrons in the filament move back and forth very quickly. These moving electrons do work by heating up the filament and making it glow.

15. The work done to move one electron in a light bulb is very small. Write down two reasons why the power is large. (**Hints:** How many electrons are in the filament of a light bulb? How did you use more power in your second trial?)

_____

_____

_____

_____

_____

_____

A Powerful Workout, continued

Communicate Results

16. Your teacher will provide a class data table on the board. Write your average power in the table. Calculate the average power for the class. How many light bulbs would it take to equal the power of one student?

_____

_____

_____

_____

_____

_____

_____

_____

_____

DATASHEET

28   STUDENT WORKSHEET

Inclined to Move

In this lab, you will examine a simple machine—an inclined plane. Your task is to compare the work done with and without the inclined plane and to analyze the effects of friction.

MATERIALS

- string
- small book
- spring scale
- meterstick
- wooden board
- blocks
- graph paper

Collect Data

1. You will use the table below to record your data.

Force vs. Height

Ramp height (cm)	Output force (N)	Ramp length (cm)	Input force (N)
10			
20			
30			
40			
50			

2. Tie a piece of string around a book. Attach the spring scale to the string. Use the spring scale to slowly lift the book to a height of 50 cm. Record the output force (the force needed to lift the book) on the line below. The output force is constant throughout the lab.

_____

3. Create a ramp using the board and blocks. The ramp should be 10 cm high at the highest point. Measure and record the ramp length on the line below.

_____

4. Keeping the spring scale and string parallel to the ramp, as shown on page 664 of the textbook, slowly pull the book up the ramp. Record the input force (the force needed to pull the book up the ramp) in the table above.

5. Increase the height of the ramp by 10 cm. Repeat step 4. Repeat this step for each ramp height up to 50 cm.

Analyze the Results

6. The *real* amount of work done includes the work done to overcome friction. Calculate the real work at each height by multiplying the ramp length by the input force (remember to convert centimeters to meters).

Graph your results, plotting real work (*y*-axis) versus height (*x*-axis).

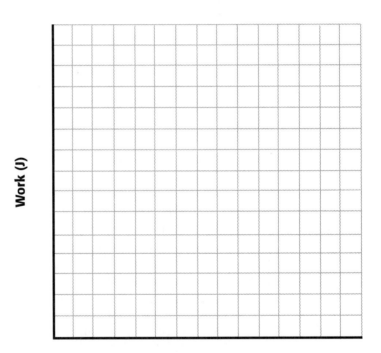

Height (cm)

7. The *ideal* amount of work is the work you would do if there were no friction. Calculate the ideal work at each height by multiplying the ramp height (m) by the output force. Plot the data on your graph.

Inclined to Move, continued

Draw Conclusions

8. Does it require more or less force to raise the book using the ramp? Explain, using your calculations and graphs.

_____

_____

Does it require more or less work to raise it using the ramp? Explain.

_____

_____

_____

_____

9. What is the relationship between the height of the inclined plane and the input force?

_____

_____

10. Write a statement that summarizes why the slopes of the two graphs are different.

_____

_____

_____

_____

_____

_____

_____

_____

CHAPTER 8

Name _____ Date _____ Class_____

DESIGN YOUR OWN

Building Machines

You are surrounded by machines. Some are simple machines, such as ramps for wheelchair access to a building. Others are compound machines, like elevators and escalators, that are made of two or more simple machines. In this lab, you will design and build several simple machines and a compound machine.

MATERIALS

- bottle caps
- cardboard
- craft sticks
- empty thread spools
- glue
- modeling clay
- paper
- pencils
- rubber bands
- scissors
- shoe boxes
- stones
- straws
- string
- tape
- other materials available in your classroom that are approved by your teacher

Procedure

1. Using any of the listed materials, build a model of each simple machine: inclined plane, lever, wheel and axle, pulley, screw, and wedge. Describe and draw each of your machines below.

Building Machines, continued

2. Design a compound machine using the materials listed. You may design a machine that already exists, or you may invent your own machine—be creative! Describe and draw your machine below.

3. After your teacher approves your design, build your compound machine.

Analyze the Results

4. List a possible use for each of your simple machines.

_____

_____

_____

_____

_____

5. Compare your simple machines with those created by your classmates. How are they different? How are they similar?

_____

_____

_____

_____

_____

CHAPTER 8

Building Machines, continued

6. How many simple machines are in your compound machine? List them.

_____

_____

_____

_____

_____

7. Compare your compound machine with those created by your classmates. How is yours different? How is yours similar?

_____

_____

_____

_____

8. What is a possible use for your compound machine? Why did you design it as you did?

_____

_____

_____

_____

9. A compound machine is listed in the Materials list. What is it?

_____

_____

Going Further

Design a compound machine that has all the simple machines in it. Explain what the machine will do and how it will make work easier. With your teacher's approval, build your machine.

DATASHEET

30 STUDENT WORKSHEET

Wheeling and Dealing

A wheel and axle is one type of simple machine. A crank handle, such as that used in pencil sharpeners, ice-cream makers, and water wells, is one kind of wheel and axle. In this lab, you will use a crank handle to find out how a wheel and axle helps you do work. You will also determine what effect the length of the handle has on the operation of the machine.

MATERIALS

- wheel and axle assembly
- meterstick
- large mass
- spring scale
- handles
- 0.5 m string
- 2 C-clamps

Collect Data

1. Use the table below to record your data.

2. Measure the radius (in meters) of the large dowel in the wheel and axle assembly. This is the axle radius, which remains constant throughout the lab. Record the axle radius in Table 1. (Hint: Measure the diameter and divide by two.)

3. Using the spring scale, measure the weight of the large mass. This is the output force, and it remains constant throughout the lab. Record the output force in Table 1.

4. Use the two C-clamps to secure the wheel-and-axle assembly to the table, as shown on page 666 of your textbook.

5. Measure the length (in meters) of handle 1. This is the radius of the wheel. Record the wheel radius in Table 1.

Table 1: Data Collection

Handle	Axle radius (m)	Output force (N)	Wheel radius (m)	Input force (N)
1				
2				
3				
4				

6. Insert the handle into the hole in the axle. Attach one end of the string to the large mass and the other end to the screw in the axle. The mass should hang down and the handle should turn freely.

CHAPTER 8

Wheeling and Dealing, continued

7. Turn the handle several times to lift the mass off the floor. Hold the spring scale upside down, and attach it to the end of the handle. Measure the force (in newtons) as the handle pulls up on the spring scale. Record this as the input force.

8. Remove the spring scale, and lower the mass to the floor. Remove the handle.

9. Repeat steps 5 through 8 with the other three handles. Record all data in Table 1.

Analyze the Results

10. You will record your calculations in the table below.

Table 2: Calculations

Handle	Axle distance (m)	Wheel distance (m)	Work input (J)	Work output (J)	Mechanical efficiency	Mechanical advantage (%)
1						
2						
3						
4						

11. a. Calculate the distance the axle rotates and the distance the wheel rotates for each handle using the equations below. Record your answers in Table 2.

Distance axle rotates = $2 \times \pi \times$ axle radius

Distance wheel rotates = $2 \times \pi \times$ wheel radius

(Use 3.14 for the value of π.) Record your answers in Table 2.

Wheeling and Dealing, continued

b. Calculate the work input and the work output using the equations below. Record your answers in Table 2.

Work input = input force × wheel distance

Work output = output force × axle distance

c. Calculate the mechanical efficiency of the wheel and axle using the equation below. Record your answers in Table 2.

$$\text{Mechanical efficiency} = \frac{\text{work output}}{\text{work input}} \times 100$$

d. Calculate the mechanical advantage of the wheel and axle using the equation below. Record your answers in Table 2.

$$\text{Mechanical advantage} = \frac{\text{wheel radius}}{\text{axle radius}}$$

CHAPTER 8

Draw Conclusions

12. What happens to work output as the handle length increases? Why?

_____

_____

What happens to work input as the handle length increases? Why?

_____

_____

_____

Name _____ Date _____ Class_____

13. What happens to mechanical efficiency as the handle length increases? Why?

_____

_____

_____

_____

_____

_____

14. What happens to mechanical advantage as the handle length increases? Why?

_____

_____

_____

_____

_____

15. What will happen to mechanical advantage if the handle length is kept constant and the axle radius gets larger?

_____

_____

_____

16. What factors were controlled in this experiment?

_____

_____

_____

What was the variable?

_____

_____

_____

DATASHEET

31 **STUDENT WORKSHEET**

Finding Energy

When you coast down a big hill on a bike or skateboard, you may notice that you pick up speed. Because you are moving, you have kinetic energy—the energy of motion. Where does that energy come from? In this lab you will find out!

MATERIALS
• 2 or 3 books
• wooden board
• masking tape
• meterstick
• metric balance
• rolling cart
• stopwatch

SCIENTIFIC METHOD

Form a Hypothesis

1. Where does the kinetic energy come from when you roll down a hill? Record your hypothesis.

_____

_____

_____

Conduct an Experiment

2. Use the table below to record your data.

Table 1: Data Collection

Height of ramp (m)	Length of ramp (m)	Mass of cart (kg)	Weight of cart (N)	Time of trial (s)			Average time (s)
				1	2	3	

3. Make a ramp with the books and board.

4. Use masking tape to make a starting line. Be sure the starting line is far enough from the top so the cart can be placed behind the line.

5. Place a strip of masking tape at the bottom of the ramp to mark the finish line.

6. Determine the height of the ramp by measuring the height of the starting line and subtracting the height of the finish line. Record the height of the ramp in meters in Table 1.

7. Measure the distance in meters between the starting and the finish lines. Record this distance as the length of the ramp in Table 1.

CHAPTER 9 ▲▲▲

Finding Energy, continued

8. Use the metric balance to find the mass of the cart in grams. Convert this to kilograms by dividing by 1,000. Record the mass in kilograms in Table 1.

9. Multiply the mass by 10 to get the weight of the cart in newtons. Record the weight in Table 1.

Collect Data

10. Set the cart behind the starting line, and release it. Use the stopwatch to time how long it takes for the cart to reach the finish line. Record the time in Table 1.

11. Repeat step 10 twice more, and average the results. Record the average time in Table 1.

Analyze the Results

12. You will record your calculations in the table below.

Table 2: Calculations

Average speed (m/s)	Final speed (m/s)	Kinetic energy at bottom (J)	Gravitational potential energy at top (J)

13. **a.** Calculate the average speed of the cart using the equation below. Record this value in Table 2. Be sure to show your work.

$$\text{Average speed} = \frac{\text{length of ramp}}{\text{average time}}$$

b. Calculate the final speed of the cart using the equation below. (This equation works because the cart accelerates smoothly from 0 m/s.) Record this value in Table 2. Be sure to show your work.

Full speed = 2 × average speed

Finding Energy, continued

    c. Calculate the kinetic energy of the cart using the equation below. Record this value in Table 2. Be sure to show your work. (Remember that $1 \text{ kg} \bullet \text{m}^2/\text{s}/\text{s} = 1 \text{ J}$, the unit used to express energy.)

$$\text{Kinetic energy} = \frac{(\text{mass}) \times (\text{final speed})^2}{2}$$

    d. Calculate the potential energy of the cart using the equation below. Record this value in Table 2. Be sure to show your work. (Remember that $1 \text{ N} = 1 \text{ kg} \bullet \text{m}/\text{s}/\text{s}$, so $1 \text{ N} \times 1 \text{ m} = 1 \text{ kg} \bullet \text{m}^2/\text{s}/\text{s} = 1 \text{ J}$.)

    Gravitational potential energy = weight × height

Draw Conclusions

14. How does the cart's gravitational potential energy at the top of the ramp compare with its kinetic energy at the bottom?

_____

_____

_____

_____

Does this support your hypothesis? Explain your answer.

_____

_____

_____

_____

▲ CHAPTER 9
▲
▲

15. You probably found that the gravitational potential energy of the cart at the top of the ramp was close but not exactly equal to the kinetic energy of the cart at the bottom. Explain this finding.

_____

_____

_____

_____

_____

_____

_____

16. While riding your bike, you coast down both a small hill and a large hill. Compare your final speed at the bottom of the small hill with your final speed at the bottom of the large hill. Explain your answer.

_____

_____

_____

_____

_____

_____

_____

_____

_____

DATASHEET

32 **STUDENT WORKSHEET**

Energy of a Pendulum

A pendulum clock is a compound machine that uses stored energy to do work. A spring is used to store energy, and with each swing of the pendulum, some of that stored energy is used to move the hands of the clock. In this lab you will take a close look at the energy conversions that occur as a pendulum swings.

MATERIALS
• 1 m of string
• 100 g hooked mass
• marker
• meterstick

Collect Data

1. Make a pendulum by tying the string around the hook of the mass. Use the marker and the meterstick to mark points on the string that are 50 cm, 70 cm, and 90 cm away from the mass.

2. Hold the string at the 50 cm mark. Gently pull the mass to the side, and release it without pushing it. Observe at least 10 swings of the pendulum.

3. Record your observations below. Be sure to note how fast and how high the pendulum swings.

_____

_____

_____

_____

_____

4. Repeat steps 2 and 3 while holding the string at the 70 cm mark and again while holding the string at the 90 cm mark. Record your observations.

_____

_____

_____

_____

_____

CHAPTER 9

Analyze the Results

5. List similarities and differences in the motion of the pendulum during all three trials.

_____

_____

_____

_____

6. At which point (or points) was the pendulum moving the slowest?

_____

_____

_____

_____

At which point (or points) of the swing was the pendulum moving the fastest?

_____

_____

_____

Draw Conclusions

7. In each trial, at which point (or points) of the swing did the pendulum have the greatest potential energy? (**Hint:** Think about your answers to question 6.)

_____

_____

_____

At which point (or points) of the swing did the pendulum have the least potential energy?

_____

_____

_____

Energy of a Pendulum, continued

8. At which point (or points) of the swing did the pendulum have the greatest kinetic energy? Explain your answers.

_____

_____

_____

At which point (or points) of the swing did the pendulum have the least kinetic energy? Explain your answers.

_____

_____

_____

9. Describe the relationship between the pendulum's potential energy and its kinetic energy on the pendulum's downward swing.

_____

_____

_____

10. What improvements might reduce the amount of energy used to overcome friction so that the pendulum would swing for a longer period of time?

_____

_____

_____

_____

_____

_____

_____

CHAPTER 9

DATASHEET

33 **STUDENT WORKSHEET**

DESIGN YOUR OWN

Eggstremely Fragile

All moving objects have kinetic energy. The faster an object is moving, the more kinetic energy it has. When a falling object hits the floor, the law of conservation of energy requires that the energy be transferred to another object or changed into another form of energy.

When an unprotected egg hits the ground from a height of 1 m, most of the kinetic energy of the falling egg is transferred to the pieces of the shell—with messy results. In this lab you will design a protection system for an egg.

MATERIALS

- raw egg
- empty half-pint milk carton
- assorted materials provided by your teacher

Conduct an Experiment

1. Using the materials provided by your teacher, design a protection system that will prevent the egg from breaking when it is dropped from heights of 1, 2, and 3 m. Keep the following points in mind while developing your egg-protection system:

   a. The egg and its protective materials must fit inside the closed milk carton. (Note: The milk carton will not be dropped with the egg.)

   b. The protective materials don't have to be soft.

   c. The protective materials can surround the egg or can be attached to the egg at various points.

2. Explain why you chose the materials you did.

   _____

   _____

   _____

   _____

3. You will perform the three trials at a time and location specified by your teacher. Record your results for each trial.

   _____

   _____

   _____

   _____

Eggstremely Fragile, continued

Analyze the Results

4. Did your egg survive all three trials? If it did not, why did your egg-protection system fail? If your egg did survive, what features of your egg-protection system transferred or absorbed the energy?

_____

_____

_____

_____

_____

_____

_____

_____

_____

_____

Draw Conclusions

5. How do egg cartons like those you find in a grocery store protect eggs from mishandling?

_____

_____

_____

_____

_____

CHAPTER 9

DATASHEET

34 **STUDENT WORKSHEET**

DISCOVERY LAB

Feel the Heat

Heat is the transfer of energy between objects at different temperatures. Energy moves from objects at higher temperatures to objects at lower temperatures. If two objects are left in contact for a while, the warmer object will cool down, and the cooler object will warm up until they eventually reach the same temperature. In this activity, you will combine equal masses of water and iron nails at different temperatures to determine which has a greater effect on the final temperature.

MATERIALS

- rubber band
- 10–12 nails
- metric balance
- 30 cm of string
- 9 oz plastic-foam cups (2)
- hot water
- 100 mL graduated cylinder
- cold water
- thermometer
- paper towels

SCIENTIFIC **METHOD**

Make a Prediction

1. When you combine substances at two different temperatures, will the final temperature be closer to the initial temperature of the warmer substance, the colder substance, or halfway in between?

_____

Conduct an Experiment/Collect Data

2. Examine the table below.

3. Use the rubber band to bundle the nails together. Find and record the mass of the bundle. Tie a length of string around the bundle, leaving one end of the string 15 cm long.

Data Collection Table

Trial	Mass of nails (g)	Volume of water that equals mass of nails (mL)	Initial temp. of water and nails (°C)	Initial temp. of water to which nails will be transferred (°C)	Final temp. of water and nails combined (°C)
1					
2					

4. Put the bundle of nails into one of the cups, letting the string dangle outside the cup. Fill the cup with enough hot water to cover the nails, and set it aside for at least 5 minutes.

5. Use the graduated cylinder to measure enough cold water to exactly equal the mass of the nails (1 mL of water = 1 g). Record this volume in the table.

6. Measure and record the temperature of the hot water with the nails and the temperature of the cold water.

7. Use the string to transfer the bundle of nails to the cup of cold water. Use the thermometer to monitor the temperature of the water-nail mixture. When the temperature stops changing, record this final temperature in the table.

_____

Feel the Heat, continued

8. Empty the cups, and dry the nails.

9. For Trial 2, repeat steps 3 through 8, but switch the hot and cold water. Record all your measurements.

Analyze the Results

10. In Trial 1, you used equal masses of cold water and nails. Did the final temperature support your initial prediction? Explain.

_____

_____

11. In Trial 2, you used equal masses of hot water and nails. Did the final temperature support your initial prediction? Explain.

_____

_____

12. In Trial 1, which material—the water or the nails—changed temperature the most after you transferred the nails? What about in Trial 2? Explain your answers.

_____

_____

_____

Draw Conclusions

13. The cold water in Trial 1 gained energy. Where did the energy come from?

_____

14. How does the energy gained by the nails in Trial 2 compare with the energy lost by the hot water in Trial 2? Explain.

_____

_____

_____

_____

_____

_____

CHAPTER 10

15. Which material seems to be able to hold energy better? Explain your answer.

_____

_____

_____

_____

_____

16. Specific heat capacity is a property of matter that indicates how much energy is required to change the temperature of 1 kg of a material by 1°C. Which material in this activity has a higher specific heat capacity (changes temperature less for the same amount of energy)?

_____

17. Would it be better to have pots and pans made from a r   ʾɑl with a high specific heat capacity or a low specific heaɪ ɑpacity? Explain your answer. (Hint: Do you want the pan or the food in the pan to absorb all the energy from the stove?)

_____

_____

_____

_____

_____

Communicate Results

18. Share your results with your classmates. Discuss how you would change your prediction to include your knowledge of specific heat capacity.

_____

_____

_____

_____

_____

DATASHEET

35 STUDENT WORKSHEET

DESIGN
YOUR OWN

Save the Cube!

The biggest enemy of an ice cube is the transfer of thermal energy, also known as heat. Energy can be transferred to an ice cube in three ways: conduction (the transfer of energy through direct contact), convection (the transfer of energy by the movement of a liquid or gas), and radiation (the transfer of energy through matter or space). Your challenge in this activity is to design a way to protect an ice cube as much as possible from all three types of energy transfer.

MATERIALS

- small plastic bag
- ice cube
- assorted materials provided by your teacher
- empty half-pint milk carton
- metric balance
- small plastic or paper cup

Procedure

1. Follow these guidelines: Use a plastic bag to hold the ice cube and any melted water. You may use any of the materials to protect the ice cube. The ice cube, bag, and protection must all fit inside the milk carton.

2. Write a description of your proposed design below. Explain how your design protects against each type of energy transfer.

_____

_____

_____

_____

_____

_____

_____

3. Find the mass of the empty cup and record it below. Then find and record the mass of an empty plastic bag.

_____

_____

4. Place an ice cube in the bag. Quickly find and record their mass together.

_____

5. Quickly wrap the bag (and the ice cube inside) in its protection. Remember that the package must fit in the milk carton.

CHAPTER 10

6. Place your protected ice cube in the "thermal zone" set up by your teacher. After 10 minutes, carefully remove the package from the thermal zone, and remove the protection from the plastic bag and ice cube.

7. Open the bag. Pour any water into the cup. Find and record the mass of the cup and water together.

_____

8. Find and record the mass of the water by subtracting the mass of the empty cup from the mass of the cup and water.

_____

9. Use the same method to find and record the mass of the ice cube.

_____

10. Find the percentage of the ice cube that melted using the following equation:

$$\% \text{ melted} = \left(\frac{\text{mass of water}}{\text{mass of ice}}\right) \times 100$$

_____

11. Record your percentage below and on the board.

_____

Analysis

12. Compared with other designs in your class, how well did your design protect against each type of energy transfer? How could you improve your design?

_____

_____

_____

13. Why is a white plastic-foam cooler so useful for keeping ice frozen?

_____

_____

_____

DATASHEET

36 **STUDENT WORKSHEET**

MAKING MODELS

Counting Calories

Energy transferred by heat is often expressed in units called calories. In this lab, you will build a model of a device called a calorimeter. Scientists often use calorimeters to measure the amount of energy that can be transferred by a substance. In this experiment, you will construct your own calorimeter and test it by measuring the energy released by a hot penny.

MATERIALS

- small plastic-foam cup with lid
- thermometer
- large plastic-foam cup
- water
- 100 mL graduated cylinder
- tongs
- heat source
- penny
- stopwatch

Procedure

1. You will use the table below to record your data.

2. Place the lid on the small plastic-foam cup, and insert a thermometer through the hole in the top of the lid. (The thermometer should not touch the bottom of the cup.) Place the small cup inside the large cup to complete the calorimeter.

3. Remove the lid from the small cup, and add 50 mL of room-temperature water to the cup. Measure the water's temperature and record the value in the 0 seconds column of the table.

4. Using tongs, carefully heat the penny. Add the penny to the water in the small cup, and replace the lid. Start your stopwatch.

5. Every 15 seconds, measure and record the temperature. Gently swirl the large cup to stir the water. Continue recording temperatures for 2 minutes (120 seconds).

Data Collection Table

Seconds	0	15	30	45	60	75	90	105	120
Water temp. (°C)									

Analysis

6. What was the total temperature change of the water after 2 minutes?

7. The number of calories absorbed by water is the mass of the water (in grams) multiplied by the temperature change (in °C) of the water. How many calories were absorbed by the water? (Hint: 1 mL of water = 1 g of water.)

8. In terms of heat, explain where the calories to change the water temperature came from.

_____

_____

_____

_____

_____

_____

Name _____ Date _____ Class _____

Made to Order

Imagine that you are a new employee at the Elements-4-U Company, which custom builds elements. Your job is to construct the atomic nucleus for each element ordered by your clients. You were hired for the position because of your knowledge about what a nucleus is made of and your understanding of how isotopes of an element differ from each other. Now it's time to put that knowledge to work!

Procedure

1. Use the table below to record your data.

Data Table

	Hydrogen-1	Hydrogen-2	Helium-3	Helium-4	Lithium-7	Beryllium-9	Beryllium-10
No. of protons							
No. of neutrons							
Atomic number							
Mass number							

MATERIALS

- 4 protons (white plastic-foam balls, 2–3 cm in diameter)
- 6 neutrons (blue plastic-foam balls, 2–3 cm in diameter)
- 20 strong-force connectors (toothpicks)
- periodic table

2. Your first assignment: the nucleus of hydrogen-1. Pick up one proton (a white plastic-foam ball). Congratulations! You have just built a hydrogen-1 nucleus, the simplest nucleus possible.

3. Count the number of protons and neutrons in the nucleus, and fill in the corresponding rows for this element in the table.

4. Determine the atomic number and mass number of the element. Record this information in the table.

5. Draw a picture of your model in the space below.

_____

Made to Order, continued

_____

6. Hydrogen-2 is an isotope of hydrogen that has one proton and one neutron. Using a strong-force connector, add a neutron to your hydrogen-1 nucleus. (Remember that in a nucleus, the protons and neutrons are held together by the strong force, which is represented in this activity by the toothpicks.) Repeat steps 3–5.

7. Helium-3 is an isotope of helium that has two protons and one neutron. Add one proton to your hydrogen-2 nucleus to create a helium-3 nucleus. Each particle should be connected to the other two particles so they make a triangle, not a line. Protons and neutrons always form the smallest arrangement possible because the strong force pulls them together. Repeat steps 3–5.

8. For the next part of the lab, you will need to use information from the periodic table of the elements. Look at the periodic table on pp. 744 and 745 of your textbook. For your job, the most important information in the periodic table is the atomic number. You can find the atomic number of any element at the top of its entry on the table. For example, the atomic number of carbon is 6.

9. Use the information in the periodic table to build models of the following isotopes of elements: helium-4, lithium-7, beryllium-9, and beryllium-10. Remember to put the protons and neutrons as close together as possible—each particle should attach to at least two others. Repeat steps 3–5 for each isotope.

Made to Order, continued

Analyze the Results

10. What is the relationship between the number of protons and the atomic number?

_____

_____

11. If you know the atomic number and the mass number of an isotope, how could you figure out the number of neutrons in its nucleus?

_____

_____

_____

12. Look up uranium on the periodic table.

    a. What is the atomic number of uranium?

    _____

    b. How many neutrons does the isotope uranium-235 have?

    _____

Communicate Results

13. Compare your model with the models of other groups. How are they similar?

_____

_____

How are they different?

_____

_____

Going Further

Working with another group, combine your models. Identify the element (and isotope) you have created.

_____

_____

DATASHEET

38 **STUDENT WORKSHEET**

MAKING MODELS

Create a Periodic Table

You probably have classification systems for many things in your life, such as your clothes, your books, and your CDs. One of the most important classification systems in science is the periodic table of the elements. In this lab you will develop your own classification system for a collection of ordinary objects. You will analyze trends in your system and compare your system with the periodic table of the elements.

MATERIALS

- bag of objects
- 20 squares of paper, each 3 × 3 cm
- metric balance
- metric ruler
- 2 sheets of graph paper

Procedure

1. Your teacher will give you a bag of objects. Your bag is missing one item. Examine the items carefully. Describe the missing object in as many ways as you can. Be sure to include the reasons why you think the missing object has these characteristics.

_____

_____

_____

_____

_____

_____

2. Lay the paper squares out on your desk or table so that you have a grid of five rows of four squares each.

3. Arrange your objects on the grid in a logical order. (You must decide what order is logical!) You should end up with one blank square for the missing object.

4. Describe the basis for your arrangement.

_____

_____

_____

_____

_____

_____

Create a Periodic Table, continued

5. Measure the mass (g) and diameter (mm) of each object, and record your results in the appropriate square. Each square (except the empty one) should have one object and two written measurements on it.

6. Examine your pattern again. Does the order in which your objects are arranged still make sense? Explain.

_____

_____

_____

_____

_____

7. Rearrange the squares and their objects if necessary to improve your arrangement. Describe the basis for the new arrangement.

_____

_____

_____

_____

_____

_____

_____

_____

_____

8. Working across the rows, number the squares 1 to 20. When you get to the end of a row, continue numbering in the first square of the next row.

9. Copy your grid into your ScienceLog. In each square, be sure to list the type of object and label all measurements with appropriate units.

Analyze the Results

10. Make a graph of mass (*y*-axis) versus object number (*x*-axis). Label each axis and put a title on the graph.

_____

Create a Periodic Table, continued
_____

11. Now make a graph of diameter (*y*-axis) versus object
number (*x*-axis).

Communicate Results

12. Discuss each graph with your classmates. Try to identify
any important features of the graph. For example, does the
graph form a line or a curve? Is there anything unusual
about the graph? What do these features tell you? Write
your answers below.

_____

_____

_____

_____

_____

_____

_____

Draw Conclusions

13. How is your arrangement of objects similar to the periodic
table of the elements found in your textbook? How is your
arrangement different from that periodic table?

_____

_____

_____

_____

_____

_____

_____

_____

_____

Create a Periodic Table, continued

_____

14. Look back at your prediction about the missing object. Do you think it is still accurate? Try to improve your description by estimating the mass and diameter of the missing object. Record your estimates.

_____

_____

_____

_____

_____

_____

_____

_____

_____

_____

15. Mendeleev created a periodic table of elements and predicted characteristics of missing elements. How is your experiment similar to Mendeleev's work?

_____

_____

_____

_____

_____

CHAPTER 12

DATASHEET

| 39 | **STUDENT WORKSHEET** |

MAKING MODELS

Covalent Marshmallows

A hydrogen atom has one electron in its outer energy level, but two electrons are required to fill its outer level. An oxygen atom has six electrons in its outer energy level, but eight electrons are required to fill its outer level. In order to fill their outer energy levels, two atoms of hydrogen and one atom of oxygen can share electrons, as shown below. Such a sharing of electrons to fill the outer level of atoms is called covalent bonding. When hydrogen and oxygen bond in this manner, a molecule of water is formed. In this lab you will build a three-dimensional model of water in order to better understand the covalent bonds formed in a water molecule.

MATERIALS

- marshmallows (2 of one color, 1 of another color)
- toothpicks

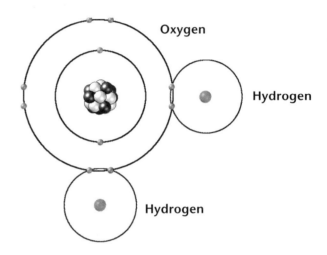

A Model of a Water Molecule

Procedure

1. Using the marshmallows and toothpicks, create a model of a water molecule. Use the diagram above for guidance in building your model.

2. Draw a sketch of your model in the space below. Be sure to label the hydrogen and oxygen atoms on your sketch.

Covalent Marshmallows, continued

---

3. In the space below, draw an electron-dot diagram of the
water molecule. (Refer to the chapter text if you need help
drawing an electron-dot diagram.)

Analysis

4. What do the marshmallows represent? What do the toothpicks represent?

_____

_____

_____

_____

_____

5. Why are the marshmallows different colors?

_____

_____

_____

_____

_____

▲ **CHAPTER 13**

Covalent Marshmallows, continued

6. Compare your model with the picture on page 102. How might your model be improved to more accurately represent a water molecule?

_____

_____

_____

_____

_____

7. Hydrogen in nature can covalently bond to form hydrogen molecules H_2. How could you model this using the marshmallows and toothpicks?

_____

_____

_____

_____

8. In the space below, draw an electron-dot diagram of an H_2 molecule.

_____

Covalent Marshmallows, continued

9. Which do you think would be more difficult to create—a model of an ionic bond or a model of a covalent bond? Explain your answer.

_____

_____

_____

_____

_____

_____

_____

_____

_____

Going Further

Create a model of a carbon dioxide molecule, which consists of two oxygen atoms and one carbon atom. The structure is similar to the structure of water, although the three atoms bond in a straight line instead of at angles. The bond between each oxygen atom and the carbon atom in a carbon dioxide molecule is a "double bond," so use two connections. Do the double bonds in carbon dioxide appear stronger or weaker than the single bonds in water? Explain your answer.

_____

_____

_____

_____

_____

_____

_____

_____

_____

DATASHEET

40 **STUDENT WORKSHEET**

Finding a Balance

Usually, balancing a chemical equation involves just writing in your ScienceLog. But in this activity, you will use models to practice balancing chemical equations, as shown below. By following the rules, you will soon become an expert equation balancer!

Example

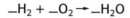

   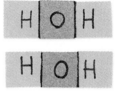

Balanced Equation

$2H_2 + O_2 \rightarrow 2H_2O$

MATERIALS
• envelopes, each labeled with an unbalanced equation

Procedure

1. The rules:

   a. Reactant-molecule models may be placed only to the left of the arrow.

   b. Product-molecule models may be placed only to the right of the arrow.

   c. You may use only complete molecule models.

   d. At least one of each of the reactant and product molecules shown in the equation must be included in the model when you are finished.

2. Select one of the labeled envelopes. In the space below, copy the unbalanced equation written on the envelope.

_____

Finding a Balance, continued

3. Open the envelope, and pull out the molecule models and the arrow. Place the arrow in the center of your work area.

4. Put one model of each molecule that is a reactant on the left side of the arrow and one model of each product on the right side.

5. Add one reactant-molecule or product-molecule model at a time until the number of each of the different-colored squares on each side of the arrow is the same. Remember to follow the rules.

6. When the equation is balanced, count the number of each of the molecule models you used. Write these numbers as coefficients, as shown in the balanced equation on page 106.

_____

_____

7. Select another envelope, and repeat the steps until you have balanced all of the equations. Record your work below.

Analysis

8. The rules specify that you are only allowed to use complete molecule models. How is this similar to what occurs in a real chemical reaction?

_____

_____

Finding a Balance, continued

9. In chemical reactions, energy is either released or absorbed. Devise a way to improve the model to show energy being released or absorbed.

_____

_____

_____

_____

_____

_____

DATASHEET

41 STUDENT WORKSHEET

DISCOVERY
LAB

Cata-what? Catalyst!

Catalysts increase the rate of a chemical reaction without being changed during the reaction. In this experiment, hydrogen peroxide, H_2O_2, decomposes into oxygen, O_2, and water, H_2O. An enzyme present in liver cells acts as a catalyst for this reaction. You will investigate the relationship between the amount of the catalyst and the rate of the decomposition reaction.

MATERIALS

- 10 mL test tubes (3)
- masking tape
- 600 mL beaker
- hot water
- funnel
- 10 mL graduated cylinder
- hydrogen peroxide solution
- 2 small liver cubes
- mortar and pestle
- tweezers

SCIENTIFIC
METHOD

Ask a Question

1. How does the amount of a catalyst affect reaction rate?

Form a Hypothesis

2. Write a statement that answers the question above. Explain your reasoning.

_____

_____

_____

Test the Hypothesis

3. Put a small piece of masking tape near the top of each test tube, and label the tubes "1," "2," and "3."

4. Create a hot-water bath by filling the beaker half-full with hot water.

5. Using the funnel and graduated cylinder, measure 5 mL of the hydrogen peroxide solution into each test tube. Place the test tubes in the hot-water bath for 5 minutes.

6. While the test tubes warm up, grind one liver cube with the mortar and pestle.

7. After 5 minutes, use the tweezers to place the cube of liver in test tube 1. Place the ground liver in test tube 2. Leave test tube 3 alone.

Make Observations

8. Observe the reaction rate (the amount of bubbling) in all three test tubes, and record your observations.

_____

_____

_____

▲ **CHAPTER 14**

Analyze the Results

9. Does liver appear to be a catalyst? Explain your answer.

_____

_____

_____

10. Which type of liver (whole or ground) produced a faster reaction? Why?

_____

_____

_____

_____

11. What is the purpose of test tube 3?

_____

_____

_____

_____

Draw Conclusions

12. How do your results support or disprove your hypothesis?

_____

_____

_____

_____

13. Why was a hot-water bath used? (Hint: Look in your book for a definition of activation energy.)

_____

_____

_____

_____

DATASHEET

42 **STUDENT WORKSHEET**

SKILL
BUILDER

Putting Elements Together

A synthesis reaction is a reaction in which two or more substances combine to form a single compound. The resulting compound has different chemical and physical properties than the substances from which it is composed. In this activity, you will synthesize, or create, copper(II) oxide from the elements copper and oxygen.

MATERIALS

- metric balance
- evaporating dish
- protective gloves
- weighing paper
- copper powder
- ring stand and ring
- wire gauze
- Bunsen burner or portable burner
- spark igniter
- tongs

Conduct an Experiment/Collect Data

1. You will record your observations in the table below.

Data Collection Table

Object	Mass (g)
Evaporating dish	
Copper powder	
Copper + evaporating dish after heating	
Copper(II) oxide	

2. Use the metric balance to measure the mass (to the nearest 0.1 g) of the empty evaporating dish. Record this mass in the table.

3. Place a piece of weighing paper on the metric balance, and measure approximately 10 g of copper powder. Record the mass (to the nearest 0.1 g) in the table. **Caution:** Wear protective gloves when working with the copper powder.

4. Use the weighing paper to place the copper powder in the evaporating dish. Spread the powder over the bottom and up the sides as much as possible. Discard the weighing paper.

5. Set up the ring stand and ring. Place the wire gauze on top of the ring. Carefully place the evaporating dish on the wire gauze.

6. Place the Bunsen burner under the ring and wire gauze. Use the spark igniter to light the Bunsen burner. **Caution:** Use extreme care when working near an open flame.

CHAPTER 14

7. Heat the evaporating dish for 10 minutes.

8. Turn off the burner, and allow the evaporating dish to cool for 10 minutes. Use tongs to remove the evaporating dish and place it on the balance to determine the mass. Record the mass in the table.

9. Determine the mass of the reaction product—copper(II) oxide—by subtracting the mass of the evaporating dish from the mass of the evaporating dish and copper powder after heating. Record this mass in the table.

Analyze the Results

10. What evidence of a chemical reaction did you observe after the copper was heated?

_____

_____

11. Explain why a change in mass occurred.

_____

_____

_____

_____

_____

12. How does the change in mass support the idea that this is a synthesis reaction?

_____

_____

_____

_____

_____

_____

_____

_____

Draw Conclusions

13. Why was powdered copper used rather than a small piece of copper? (Hint: How does surface area affect the rate of the reaction?)

_____

_____

_____

_____

14. Why was the copper heated? (Hint: Look in your book for the discussion of *activation energy*.)

_____

_____

_____

_____

15. The copper bottoms of cooking pots turn black when used. How is that similar to the results you obtained in this lab?

_____

_____

_____

_____

Going Further

Rust is iron(III) oxide—the product of a synthesis reaction between iron and oxygen. How does painting a car help prevent this type of reaction?

_____

_____

_____

_____

_____

▲▲ CHAPTER 14
▲

DATASHEET

43 **STUDENT WORKSHEET**

DISCOVERY LAB

Speed Control

The reaction rate (how fast a chemical reaction happens) is an important factor to control. Sometimes you want a reaction to take place rapidly, such as when you are removing tarnish from a metal surface. Other times you want a reaction to happen very slowly, such as when you are depending on a battery as a source of electrical energy. In this lab, you will discover how changing the surface area and concentration of the reactants affects reaction rate. In this lab, you can estimate the rate of reaction by observing how fast bubbles form.

MATERIALS

- 30 mL test tubes (6)
- 6 strips of aluminum, approximately 5 × 1 cm each
- test-tube rack
- scissors
- 2 funnels
- 10 mL graduated cylinders (2)
- acid A
- acid B

 SCIENTIFIC **METHOD**

Part A–Surface Area

Ask a Question

1. How does changing the surface area of a metal affect reaction rate?

Form a Hypothesis

2. Write a statement that answers the question above. Explain your reasoning.

_____

_____

_____

Test the Hypothesis

3. Use three identical strips of aluminum. Put one strip into a test tube. Place the test tube in the test-tube rack. **Caution:** The strips of metal may have sharp edges.

4. Carefully fold a second strip in half and then in half again. Use a textbook or other large object to flatten the folded strip as much as possible. Place the strip in a second test tube in the test-tube rack.

5. Use scissors to cut a third strip of aluminum into the smallest possible pieces. Place all of the pieces into a third test tube, and place the test tube in the test-tube rack.

6. Use a funnel and a graduated cylinder to pour 10 mL of acid A into each of the three test tubes.
   Caution: Hydro-chloric acid is corrosive. If any acid should spill on you, immediately flush the area with water and notify your teacher.

Speed Control, continued

Make Observations

7. Observe the rate of bubble formation in each test tube. Record your observations below.

_____

_____

_____

_____

_____

Analyze the Results

8. Which form of aluminum had the greatest surface area? Which had the smallest?

_____

_____

_____

_____

_____

_____

_____

9. In the three test tubes, the amount of aluminum and the amount of acid were the same. Which form of the aluminum seemed to react the fastest? Which form reacted the slowest? Explain your answers.

_____

_____

_____

_____

_____

_____

▲ CHAPTER 14

10. Do your results support the hypothesis you made in step 2? Explain.

_____

_____

_____

_____

Draw Conclusions

11. Would powdered aluminum react faster or slower than the forms of aluminum you used? Explain your answer.

_____

_____

_____

_____

_____

Part B–Concentration

Ask a Question

12. How does changing the concentration of acid affect the reaction rate?

Form a Hypothesis

13. Write a statement that answers the question above. Explain your reasoning.

_____

_____

_____

Test the Hypothesis

14. Place one of the three remaining aluminum strips in each of the three clean test tubes. (Note: Do not alter the strips.) Place the test tubes in the test-tube rack.

15. Using the second funnel and graduated cylinder, pour 10 mL of water into one of the test tubes. Pour 10 mL of acid B into the second test tube. Pour 10 mL of acid A into the third test tube.

Speed Control, continued

Make Observations

16. Observe the rate of bubble formation in the three test tubes. Record your observations.

_____

_____

_____

_____

Analyze the Results

17. In this set of test tubes, the strips of aluminum were the same, but the concentration of the acid was different. Acid A is more concentrated than acid B. Was there a difference between the test tube with water and the test tubes with acid? Which test tube formed bubbles the fastest? Explain.

_____

_____

_____

_____

18. Do your results support the hypothesis you made in step 13? Explain.

_____

_____

_____

_____

Draw Conclusions

19. Explain why spilled hydrochloric acid should be diluted with water before it is wiped up.

_____

_____

_____

_____

▲ CHAPTER 14

DATASHEET

44 STUDENT WORKSHEET

SKILL BUILDER

Cabbage Patch Indicators

Indicators are weak acids or bases that change color due to the pH of the substance to which they are added. Red cabbage contains a natural indicator that turns specific colors at specific pHs. In this lab you will extract the indicator from red cabbage and use it to determine the pH of several liquids.

MATERIALS

- distilled water
- 250 mL beaker
- red cabbage leaf
- hot plate
- beaker tongs
- masking tape
- test tubes
- test-tube rack
- eyedropper
- sample liquids provided by teacher
- litmus paper

Procedure

1. You will record your results in the table below.

Data Collection Table

Liquid	Color with indicator	pH	Effect on litmus paper
control			

2. Put on protective gloves. Pour 100 mL of distilled water into the beaker. Tear the cabbage leaf into small pieces, and place the pieces in the beaker.

3. Use the hot plate to heat the cabbage and water to boiling. Continue boiling until the water is deep blue. **Caution:** Use extreme care when working near a hot plate.

4. Use tongs to remove the beaker from the hot plate, and turn the hot plate off. Allow the solution to cool for 5–10 minutes.

5. While the solution is cooling, use masking tape and a pen to label the test tubes for each sample liquid. Label one test tube as the control. Place the tubes in the rack.

6. Use the eyedropper to place a small amount (about 5 mL) of the indicator (cabbage juice) into the test tube labeled as the control. Pour a small amount (about 5 mL) of each sample liquid into the appropriate test tubes.

Name _____ Date _____ Class _____

Cabbage Patch Indicators, continued

7. Using the eyedropper, place several drops of the indicator into each test tube and swirl gently. Record the color of each liquid in the data table. Use the chart on page 689 of your textbook to determine the pH for each sample, and then record the pH in the table.

8. Litmus paper is an indicator that turns red in an acid and blue in a base. Test each liquid with a strip of litmus paper, and record the results.

Analysis

9. What purpose does the control serve?

_____

_____

_____

_____

What is the pH of the control? Explain.

_____

_____

_____

10. What colors are associated with acids? with bases?

_____

_____

_____

_____

11. Why is red cabbage juice considered a good indicator?

_____

_____

_____

_____

_____

_____

12. Which do you think would be more useful to help identify an unknown liquid—litmus paper or red cabbage juice? Why?

_____

_____

_____

_____

_____

_____

_____

_____

Going Further

Unlike distilled water, rainwater has some carbon dioxide dissolved in it. Is rainwater acidic, basic, or neutral? To find out, place a small amount of the cabbage juice indicator (which is water-based) in a clean test tube. Use a straw to gently blow bubbles in the indicator. Continue blowing bubbles until you see a color change. What can you conclude about the pH of your "rainwater?" What is the purpose of blowing bubbles in the cabbage juice?

_____

_____

_____

_____

_____

DATASHEET

45 **STUDENT WORKSHEET**

Making Salt

A neutralization reaction between an acid and a base produces water and a salt. In this lab, you will react an acid with a base and then let the water evaporate. You will then examine what is left for properties that tell you that it is indeed a salt.

MATERIALS
• hydrochloric acid
• 100 mL graduated cylinder
• 100 mL beaker
• distilled water
• phenolphthalein solution in a dropper bottle
• sodium hydroxide
• glass stirring rod
• 2 eyedroppers
• evaporating dish
• magnifying lens

Procedure

1. Put on protective gloves. Carefully measure 25 mL of hydrochloric acid in a graduated cylinder, then pour it into the beaker. Carefully rinse the graduated cylinder with distilled water to clean out any leftover acid. **Caution:** Hydrochloric acid is corrosive. If any should spill on you, immediately flush the area with water and notify your teacher.

2. Add three drops of phenolphthalein indicator to the acid in the beaker. You will not see anything happen yet because this indicator won't show its color unless too much base is present.

3. Measure 20 mL of sodium hydroxide (base) in the graduated cylinder, and add it slowly to the beaker with the acid. Use the stirring rod to mix the substances completely. **Caution:** Sodium hydroxide is also corrosive. If any should spill on you, immediately flush the area with water and notify your teacher.

4. Use an eyedropper to add more base to the acid-base mixture in the beaker a few drops at a time. Be sure to stir the mixture after each few drops. Continue adding drops of base until the mixture remains colored after stirring.

5. Use another eyedropper to add acid to the beaker, one drop at a time, until the color just disappears after stirring.

6. Pour the mixture carefully into an evaporating dish, and place the dish where your teacher tells you to. Then allow the water to evaporate overnight.

7. The next day, examine your evaporating dish and study the crystals with a magnifying lens. Identify the color, shape, and other properties of the crystals. Record your observations below.

_____

_____

_____

_____

CHAPTER 15

Analysis

8. The equation for this reaction is:

$HCl + NaOH \rightarrow H_2O + NaCl$.

NaCl is ordinary table salt and forms very regular cubic crystals that are white. Did you find white cubic crystals?

_____

_____

9. The phenolphthalein indicator changes color in the presence of a base. Why did you add more acid in step 5 until the color disappeared?

_____

_____

_____

_____

_____

_____

Going Further

Another neutralization reaction occurs between hydrochloric acid and potassium hydroxide, KOH. The equation for this reaction is as follows:

$HCl + KOH \rightarrow H_2O + KCl$

What are the products of this neutralization reaction? How do they compare with those you discovered in this experiment?

_____

_____

DATASHEET

46 **STUDENT WORKSHEET**

MAKING MODELS

Domino Chain Reactions

Fission of uranium-235 is a process that relies on neutrons. When a uranium-235 nucleus splits into two smaller nuclei, it releases two or three neutrons that can cause neighboring nuclei to undergo fission. This can result in a nuclear chain reaction. In this lab you will build two models of nuclear chain reactions using dominoes.

MATERIALS
• 15 dominoes
• stopwatch

Conduct an Experiment

1. For the first model, set up the dominoes as shown below. Each domino should hit two dominoes in the next row when pushed over.

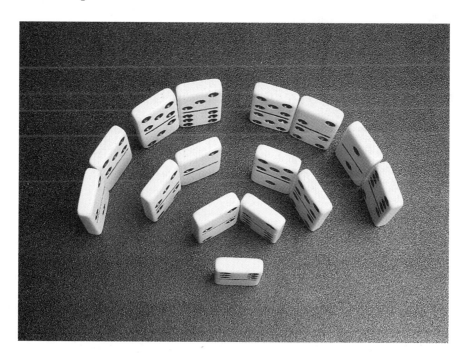

2. Measure the time it takes for all of the dominoes to fall. To do this, start the stopwatch as you tip over the front domino. Stop the stopwatch when the last domino falls.

3. If some of the dominoes do not fall, repeat steps 1 and 2. You may have to adjust the setup a few times. Record the fall time in the table at the top of the next page.

Domino Chain Reactions, continued

Data Chart for Domino Fall Times

Time for first model (in sec)	
Time for second model (in sec)	

4. For the second model, set up the dominoes as shown at left. The domino in the first row should hit both of the dominoes in the second row. Beginning with the second row, only one domino from each row should hit both of the dominoes in the next row.

5. Repeat step 2. Again, you may have to adjust the setup a few times to get all the dominoes to fall. Record the fall time in the table above.

Analyze Your Results

6. Which model represents an uncontrolled chain reaction? Explain.

_____

_____

_____

_____

Which represents a controlled chain reaction? Explain.

_____

_____

_____

_____

7. Imagine that each domino releases a certain amount of energy as it falls. Compare the total amount of energy released in the two models.

_____

_____

_____

8. Compare the time needed to release the energy in the models. Which model was longest?

_____

Which model was shortest?

_____

Draw Conclusions

9. In a nuclear power plant, a chain reaction is controlled by using a material that absorbs neutrons. Only enough neutrons to continue the chain reaction are allowed to produce further fission of uranium-235. Explain how your model of a controlled nuclear chain reaction modeled this process.

_____

_____

_____

_____

_____

_____

_____

10. Why must uranium nuclei be close to each other in order for a nuclear chain reaction to occur? (Hint: What would happen in your model if the dominoes were too far apart?)

_____

_____

_____

_____

_____

_____

DATASHEET

47 STUDENT WORKSHEET

DISCOVERY LAB

Stop the Static Electricity!

Imagine this scenario: Some of your clothes cling together when they come out of the dryer. This annoying problem is caused by static electricity—the buildup of electric charges on an object. In this lab, you'll discover how this buildup occurs.

MATERIALS
• 30 cm thread
• plastic-foam packing peanut
• tape
• rubber rod
• wool cloth
• glass rod
• silk cloth

SCIENTIFIC METHOD

Ask a Question

1. How do electric charges build up on clothes in a dryer?

Form a Hypothesis

2. Write a statement that answers the question above. Explain your reasoning.

_____

_____

_____

Test the Hypothesis

3. Tie a piece of thread approximately 30 cm in length to a packing peanut. Hang the peanut by the thread from the edge of a table. Tape the thread to the table.

4. Rub the rubber rod with the wool cloth for 10–15 seconds. Bring the rod near, but do not touch the peanut. Observe the peanut and record your observations. If nothing happens, repeat this step.

_____

_____

_____

_____

5. Touch the peanut with the rubber rod. Pull the rod away from the peanut, and then bring it near again. Record your observations.

_____

_____

_____

Stop the Static Electricity! continued

6. Repeat steps 4 and 5 with the glass rod and silk cloth.

_____

_____

_____

_____

_____

7. Now rub the rubber rod with the wool cloth, and bring the rod near the peanut again. Record your observations.

_____

_____

_____

_____

Analyze the Results

8. What caused the peanut to act differently in steps 4 and 5?

_____

_____

_____

_____

_____

_____

9. Did the glass rod have the same effect on the peanut as the rubber rod did? Explain how the peanut reacted in each case.

_____

_____

_____

_____

_____

Stop the Static Electricity! continued

10. Was the reaction of the peanut the same in steps 5 and 7?
Explain.

_____

_____

_____

_____

_____

_____

Draw Conclusions

11. Based on your results, was your hypothesis correct? Explain
your answer, and write a new statement if necessary.

_____

_____

_____

12. Explain why the rubber rod and the glass rod affected the
peanut.

_____

_____

_____

_____

_____

_____

Going Further

Do some research to find out how a dryer sheet helps stop the
buildup of electric charges in the dryer.

DATASHEET

48 **STUDENT WORKSHEET**

MAKING MODELS

Potato Power

Have you ever wanted to look inside a D cell from a flashlight or an AA cell from a portable radio? All cells include the same basic components. There is a metal "bucket," some electrolyte (a paste), and a rod of some other metal (or solid) in the middle. Even though the construction is simple, companies that manufacture cells are always trying to make a product with the highest voltage possible from the least expensive materials. Sometimes they try different pastes, and sometimes they try different combinations of metals. In this lab, you will make your own cell. Using inexpensive materials, you will try to produce the highest voltage you can.

MATERIALS
• labeled metal strips
• potato
• metric ruler
• voltmeter

Procedure

1. Choose two metal strips. Carefully push one of the strips into the potato at least 2 cm deep. Insert the second strip the same way, and measure how far apart the two strips are. (If one of your metal strips is too soft to push into the potato, push a harder strip in first, remove it, and then push the soft strip into the slit.) Record the two metals you have used and the distance between them. **Caution:** The strips of metal may have sharp edges.

   _____

   _____

   _____

2. Connect the voltmeter to the two strips, and record the voltage.

   _____

   _____

3. Move one of the strips closer to or farther from the other. Measure the new distance and voltage. Record your results.

   _____

   _____

   _____

   _____

4. Repeat steps 1 through 3 using different combinations of metal strips and distances until you find the combination that produces the highest voltage.

CHAPTER 17

Potato Power, continued

Analysis

5. What combination of metals and distance produced the highest voltage?

_____

_____

_____

6. If you change only the distance but use the same metal strips, what is the effect on the voltage?

_____

_____

_____

7. One of the metal strips tends to lose electrons, while the other tends to gain electrons. What do you think would happen if you used two strips of the same metal?

_____

_____

_____

_____

_____

DATASHEET

49 STUDENT WORKSHEET

SKILL BUILDER

Circuitry 101

You have learned that there are two basic types of electrical circuits. A series circuit connects all the parts in a single loop, and a parallel circuit connects each of the parts on separate branches to the power source. If you want to control the whole circuit, the loads and the switch must be wired in series. If you want parts of the circuit to operate independently, the loads must be wired in parallel.

No matter how simple or complicated a circuit may be, Ohm's law (current equals voltage divided by resistance) applies. In this lab, you will construct both a series circuit and a parallel circuit. You will use an ammeter to measure current and a voltmeter to measure voltage. With each circuit, you will test and apply Ohm's law.

MATERIALS
• power source—dry cell(s) • switch • 3 light-bulb holders • 3 light bulbs • insulated wire, cut into 15 cm lengths with both ends stripped • ammeter • voltmeter

Part A–Series Circuit

Procedure

1. Construct a series circuit with a power source, a switch, and three light bulbs. **Caution:** Always leave the switch open when constructing or changing the circuit. Close the switch only when you are testing or taking a reading.

2. Draw a diagram of your circuit in your ScienceLog.

3. Test your circuit. Do all three bulbs light up? Are they all the same brightness? What happens if you carefully unscrew one light bulb? Does it make any difference which bulb you unscrew? Record your observations.

_____

_____

_____

_____

_____

_____

_____

4. Connect the ammeter between the power source and the switch. Close the switch, and record the current with a label on your diagram. Be sure to show where you measured the current and what the value was.

5. Reconnect the circuit so the ammeter is between the first and second bulbs. Record the current, as you did in step 4.

6. Move the ammeter so it is between the second and third bulbs, and record the current again.

CHAPTER 17

Circuitry 101, continued

7. Remove the ammeter from the circuit, and connect the voltmeter to the two ends of the power source. Record the voltage with a label on your diagram.

8. Use the voltmeter to measure the voltage across each bulb. Label the voltage across each bulb on your diagram.

Part B–Parallel Circuit

Procedure

9. Take apart your series circuit, and reassemble the same power source, switch, and three light bulbs so that the bulbs are wired in parallel. (Note: The switch must remain in series with the power source to be able to control the whole circuit.)

10. Draw a diagram of your parallel circuit in your ScienceLog.

11. Test your circuit, and record your observations as you did in step 3.

_____

_____

_____

_____

_____

_____

_____

12. Connect the ammeter between the power source and the switch. Record the reading on your diagram.

13. Reconnect the circuit so that the ammeter is right next to one of the three bulbs. Record the current on your diagram.

14. Repeat step 13 for the two remaining bulbs.

15. Remove the ammeter from your circuit, and connect the voltmeter to the two ends of the power source. Record this voltage on your diagram.

16. Measure the voltage across each light bulb. Record the value on your diagram.

Circuitry 101, continued

Analysis—Parts A and B

17. Was the current the same at all places in the series circuit? Was it the same everywhere in the parallel circuit?

_____

_____

_____

_____

18. For each circuit, compare the voltage at each light bulb with the power source.

_____

_____

_____

_____

_____

_____

_____

19. What is the relationship between the voltage at the power source and the voltages at the light bulbs in a series circuit?

_____

_____

_____

_____

20. Use Ohm's law and the readings for current (I) and voltage (V) at the power source for both circuits to calculate the total resistance (R) in both the series and parallel circuits. Show your work below.

_____
Circuitry 101, continued

21. Was the total resistance for both circuits the same? Explain your answer.

_____

_____

_____

_____

_____

_____

22. Why did the bulbs differ in brightness?

_____

_____

_____

23. Based on your results, what do you think might happen if too many electric appliances are plugged into the same series circuit? the same parallel circuit?

_____

_____

_____

_____

_____

_____

_____

_____

_____

_____

DATASHEET

50 STUDENT WORKSHEET

Magnetic Mystery

Every magnet is surrounded by a magnetic field. Magnetic field lines show the shape of the magnetic field. These lines can be modeled by using iron filings. The iron filings are affected by the magnetic field, and they fall into lines showing the field. In this lab, you will first learn about magnetic fields, and then you will use this knowledge to identify a mystery magnet's shape and orientation based on observations of the field lines.

MATERIALS

- 2 magnets, different shapes
- sheet of clear acetate
- iron filings
- shoe box
- masking tape

Collect Data

1. Lay one of the magnets flat on a table.

2. Place a sheet of clear acetate over the magnet. Sprinkle some iron filings on the acetate to see the magnetic field lines.

3. Draw the magnet and the magnetic field lines.

4. Remove the acetate, and pour the iron filings back into the container.

5. Place your magnet so that one end is pointing up. Repeat steps 2 through 4.

6. Place your magnet on its side. Repeat steps 2 through 4.

7. Repeat steps 1 through 6 with the other magnet. Draw the magnet and field lines in the spaces above.

Conduct an Experiment

8. Create a magnetic mystery for another lab team by removing the lid from a shoe box and taping a magnet underneath the lid. Orient the magnet so that determining the shape of the magnetic field and the orientation of the magnet will be challenging. Once the magnet is secure, replace the lid on the box.

9. Exchange boxes with another team.

10. Without opening the box, use the sheet of acetate and the iron filings to determine the shape of the magnetic field of the magnet in the box.

11. Make a drawing of the magnetic field lines.

Draw Conclusions

12. Use your drawings from steps 1 through 7 to identify the shape and orientation of the magnet in your magnetic mystery box. Draw a picture of your conclusion in the space provided.

Going Further

Examine your drawings. Can you identify the north and south poles of a magnet from the shape of the magnetic field lines? Design a procedure that would allow you to determine the poles of a magnet.

_____

_____

_____

_____

Name _____ Date _____ Class _____

DISCOVERY LAB

Electricity from Magnetism

You use electricity every day. But did you ever wonder where it comes from? Some of the electrical energy you use is converted from chemical energy in cells or batteries. But what about when you plug a lamp into a wall outlet? In this lab, you will see how electricity can be generated from magnetism.

SCIENTIFIC **METHOD**

MATERIALS

- sandpaper
- 150 cm of magnet wire
- cardboard tube
- commercial galvanometer
- 2 insulated wires with alligator clips, each wire approximately 30 cm long
- strong bar magnet

Ask a Question

1. How can electricity be generated from magnetism?

Form a Hypothesis

2. Write a statement to answer the question above.

_____

_____

_____

_____

Test the Hypothesis

3. Sand the enamel off the last 2 or 3 cm of each end of the magnet wire. Wrap the magnet wire around the tube to make a coil. Attach the bare ends of the wire to the galvanometer using the insulated wires.

4. While watching the galvanometer, move a bar magnet into the coil, hold it there for a moment, and then remove it. Record your observations.

_____

_____

_____

_____

CHAPTER 18 ▲ ▲ ▲

5. Repeat step 4 several times, moving the magnet at different speeds. Observe the galvanometer carefully.

_____

_____

_____

_____

6. Hold the magnet still, and pass the coil over the magnet. Record your observations.

_____

_____

_____

_____

Analyze the Results

7. How does the speed of the magnet affect the size of the electric current?

_____

_____

8. How is the direction of the electric current affected by the motion of the magnet?

_____

_____

9. Examine your hypothesis. Is your hypothesis accurate? Explain.

_____

_____

_____

_____ **Electricity from Magnetism, continued** _____

If necessary, write a new hypothesis to answer the question in step 1.

_____

_____

_____

_____

Draw Conclusions

10. Would an electric current still be generated if the wire were broken? Why or why not?

_____

_____

11. Could a stationary magnet be used to generate an electric current? Explain.

_____

_____

12. What energy conversions occur in this investigation?

_____

_____

Communicate Results

13. Write a short explanation of the requirements for generating electricity from magnetism.

_____

_____

_____

_____

CHAPTER 18

▲
▲▲
▲

DATASHEET

52 STUDENT WORKSHEET

MAKING MODELS

Build a DC Motor

Electric motors can be used for many things. Hair dryers, CD players, and even some cars and buses are powered by electric motors. In this lab, you will build a direct current electric motor—the basis for the electric motors you use every day.

MATERIALS

- 100 cm of magnet wire
- cardboard tube
- sandpaper
- 2 large paper clips
- 4 disk magnets
- plastic-foam cup
- tape
- 2 insulated wires with alligator clips, each approximately 30 cm long
- 4.5 V battery
- permanent marker

Procedure

1. To make the armature for the motor, wind the magnet wire around the cardboard tube to make a coil like the one shown on page 700 of your textbook. Wind the ends of the wire around the loops on each side of the coil. Leave about 5 cm free on each end.

2. Hold the coil on its edge. Sand the enamel from only the top half of each free end of the magnet wire. This acts like a commutator, except that it blocks the electric current instead of reversing it during half of each rotation.

3. Partially unfold the two paper clips from the middle. Make a hook in one end of each paper clip to hold the coil.

4. Place two disk magnets in the bottom of the cup, and place the other magnets on the outside of the bottom of the cup. The magnets should remain in place when the cup is turned upside down.

5. Tape the paper clips to the sides of the cup. The hooks should be at the same height and should keep the coil from hitting the magnet.

6. Test your coil. Flick the top of the coil lightly with your finger. The coil should spin freely without wobbling or sliding to one side.

7. Make adjustments to the ends of the magnet wire and the hooks until your coil spins freely.

8. Use the alligator clips to attach one insulated wire to each paper clip.

9. Attach the free end of one insulated wire to one terminal of the battery.

Collect Data

10. Connect the free end of the other insulated wire to the second battery terminal, and give your coil a gentle spin. Record your observations.

_____

_____

Build a DC Motor, continued

11. Stop the coil and give it a gentle spin in the opposite direction. Record your observations.

_____

_____

12. If the coil does not keep spinning, check the ends of the wire. Bare wire should touch the paper clips during half of the spin, and only enamel should touch the paper clips for the other half of the spin.

13. If you removed too much enamel, color half of the exposed wire with a permanent marker.

14. Switch the connections to the battery and repeat steps 10 and 11. Record your observations.

_____

_____

_____

_____

Analyze the Results

15. Did your motor always spin in the direction you started it? Explain.

_____

_____

_____

16. Why was the motor affected by switching the battery connections?

_____

_____

_____

_____

▲ ▲ CHAPTER 18
▲

Build a DC Motor, continued

17. Some electric cars run on solar power. Which part of your model would be replaced by solar panels?

_____

Draw Conclusions

18. Some people claim that electric-powered cars are cleaner than gasoline-powered cars. Explain why this might be true.

_____

_____

19. List some reasons that electric cars are not ideal. (Hint: What happens to batteries?)

_____

_____

_____

_____

_____

20. How could your model be used to help design a hair dryer?

_____

_____

_____

_____

_____

21. Make a list of at least three other items that could be powered by an electric motor like the one you built.

_____

_____

_____

_____

DATASHEET

53 STUDENT WORKSHEET

Tune In!

You probably have listened to radios many times in your life. Modern radios are complicated electronic devices. However, radios do not have to be so complicated. The basic parts of all radios include: a diode, an inductor, a capacitor, an antenna, a ground wire, and an earphone (or a speaker and amplifier on a large radio). In this activity, you will examine each of these components one at a time as you build a working model of a radio-wave receiver.

MATERIALS

- diode
- 2 m of insulated wire
- 2 cardboard tubes
- tape
- scissors
- aluminum foil
- sheet of paper
- 7 connecting wires, 30 cm each
- 3 paper clips
- cardboard, 20 × 30 cm
- antenna
- ground wire
- earphone

Procedure

1. Examine and describe the diode.

_____

_____

_____

_____

2. A diode carries current in only one direction. Draw the inside of a diode, and illustrate how this might occur.

3. An inductor controls the amount of electric current by increasing the resistance of the wire. Make an inductor by winding the insulated wire around a cardboard tube approximately 100 times. Wind the wire so that all the turns of the coil are neat and in an orderly row. Leave about 25 cm of wire on each end of the coil. The coil of wire may be held on the tube using tape.

4. Now you will construct the variable capacitor. A capacitor stores electrical energy when an electric current is applied. A variable capacitor is a capacitor in which the amount of energy stored can be changed. Cut a piece of aluminum foil that is long enough to wrap around the second tube but is only half the length of the tube. Keep the foil as wrinkle-free as possible as you wrap it around the tube, and tape the foil to itself. Now tape the foil to the tube.

CHAPTER 19

Tune In! continued

5. Use the sheet of paper and tape to make a sliding cover on the tube. The paper should be about 1 cm longer than the foil.

6. Cut another sheet of aluminum foil to wrap around the paper. Leave approximately 1 cm of paper showing at each end of the foil. Tape this foil sheet to the paper sleeve. If you have done this correctly, you have a paper/foil sheet which will slide up and down the tube over the stationary foil. The two pieces of foil should not touch.

7. Stand your variable capacitor on its end so that the stationary foil is at the bottom. The amount of stored energy is greater when the sleeve is down than when the sleeve is up.

8. Use tape to attach one connecting wire to the stationary foil at the end of the tube. Use tape to attach another connecting wire to the sliding foil sleeve. Be sure that the metal part of the wire touches the foil.

9. Hook three paper clips on one edge of the cardboard, as shown on page 703 of your textbook. Label one paper clip A, another B, and the third C.

10. Lay the inductor on the piece of cardboard, and tape it to the cardboard.

11. Stand the capacitor next to the inductor, and tape the tube to the cardboard. Be sure not to tape the sleeve—it must be free to slide.

12. Use tape to connect the diode to paper clips A and B. The cathode should be closest to paper clip B. (The cathode end of the diode is the one with the dark band.) Make sure that all connections have good metal-to-metal contact.

13. Connect one end of the inductor to paper clip A, and the other end to paper clip C. Use tape to hold the wires in place.

14. Connect the wire from the sliding part of the capacitor to paper clip A. Connect the other wire (from the stationary foil) to paper clip C.

15. The antenna receives radio waves transmitted by a radio station. Tape a connecting wire to your antenna. Then connect this wire to paper clip A.

16. Use tape to connect one end of the ground wire to paper clip C. The other end of the ground wire should be connected to an object specified by your teacher.

17. The earphone will allow you to detect the radio waves you receive. Connect one wire from the earphone to paper clip B and the other wire to paper clip C.

18. You are now ready to begin listening. With everything connected and with the earphone in your ear, slowly slide the paper/foil sheet of the capacitor up and down. Listen for a very faint sound. You may have to troubleshoot many of the parts to get your receiver to work. As you troubleshoot, check to be sure there is good contact between all the connections.

Analysis

19. Describe the process of operating your receiver.

_____

_____

_____

_____

_____

20. Considering what you have learned about a diode, why is it important to have the diode connected correctly?

_____

_____

_____

21. A function of the inductor on a radio is to "slow the current down." Why does the inductor you made slow the current down more than a straight wire the length of your coil does?

_____

_____

_____

▲▲ **CHAPTER 19**
▲

22. A capacitor consists of any two conductors separated by an insulator. For your capacitor, list the two conductors and the insulator.

_____

_____

_____

23. Explain why the amount of stored energy is increased when you slide the foil sleeve down and decreased when the sleeve is up.

_____

_____

_____

_____

24. Make a list of ways that your receiver is similar to a modern radio.

_____

_____

_____

_____

Make a list of ways that your receiver is different from a modern radio.

_____

_____

_____

_____

_____

DATASHEET

54 **STUDENT WORKSHEET**

DISCOVERY LAB

Wave Energy and Speed

If you threw a rock into a pond, waves would carry energy away from the point of origin. But if you threw a large rock into a pond, would the waves carry more energy away from the point of origin than waves created by a small rock? And would a large rock create waves that move faster than waves created by a small rock? In this lab you'll answer these questions.

MATERIALS

- shallow pan, approximately 20 × 30 cm
- newspaper
- small beaker
- water
- 2 pencils
- stopwatch

SCIENTIFIC **METHOD**

Ask a Question

1. In this lab you will answer the following questions: Do waves created by a large disturbance carry more energy than waves created by a small disturbance? Do waves created by a large disturbance travel faster than waves created by a small disturbance?

Form a Hypothesis

2. Write a few sentences that answer the questions in step 1.

_____

_____

_____

_____

Test the Hypothesis

3. Place the pan on a few sheets of newspaper. Using the small beaker, fill the pan with water.

4. Make sure that the water is still. Tap the surface of the water near one end of the pan with the eraser end of one pencil. This represents the small disturbance. Record your observations about the size of the waves that are created and the path they take.

_____

_____

5. Repeat step 4. This time, use the stopwatch to record the amount of time it takes for one of the waves to reach the other side of the pan. Record your data.

_____

_____

CHAPTER 20

_____

Wave Energy and Speed, continued

6. Repeat steps 4 and 5 using two pencils at once. This represents the large disturbance. (Try to use the same amount of force to tap the water as you did with just one pencil.) Observe, and record your results.

_____

_____

_____

_____

_____

_____

Analyze the Results

7. Compare the appearance of the waves created by one pencil with that of the waves created by two pencils. Were there any differences in amplitude (wave height)? Record your observations.

_____

_____

_____

_____

_____

_____

8. Compare the amount of time required for the waves to reach the side of the pan. Did the waves travel faster when two pencils were used?

_____

_____

_____

_____

_____

_____

Wave Energy and Speed, continued

Draw Conclusions

9. Do waves created by a large disturbance carry more energy than waves created by a small disturbance? Explain your answer using your results to support your answer. (Hint: Remember the relationship between amplitude and energy.)

_____

_____

_____

_____

_____

10. Do waves created by a large disturbance travel faster than waves created by a small disturbance? Explain your answer. Be sure to discuss how your results support your answer.

_____

_____

_____

_____

Going Further

A tsunami is a giant ocean wave that can reach a height of 30 m. Tsunamis that reach land can cause injury and enormous property damage. Using what you learned in this lab about wave energy and speed, explain why tsunamis are so dangerous. How do you think scientists can forecast when tsunamis will reach land?

_____

_____

_____

_____

_____

_____

_____

DATASHEET

55 STUDENT WORKSHEET

Wave Speed, Frequency, and Wavelength

Wave speed, frequency, and wavelength are three related properties of waves. In this lab you will make observations and collect data to determine the relationship among these properties.

MATERIALS
• coiled spring toy
• meterstick
• stopwatch

Part A—Wave Speed

Procedure

1. Use the table below to record your data.

Table 1: Wave Speed Data

Trial	Length of spring (m)	Time for wave (s)	Speed of wave (m/s)
1			
2			
3			
Average			

2. On the floor or a table, two students should stretch the spring to a length of 2 to 4 m. A third student should measure the length of the spring. This length will be the same for all three trials. Record the length in Table 1.

3. One student should pull part of the spring sideways with one hand, as shown on page 708 of your textbook, and release the pulled-back portion. This will cause a wave to travel down the spring.

4. Using a stopwatch, the third group member should measure how long it takes for the wave to travel down the length of the spring and back. Record this time in Table 1.

5. Repeat steps 3 and 4 two more times.

Wave Speed, Frequency, and Wavelength, continued

Analyze the Results

6. Calculate and record the wave speed for each trial. (Hint: Speed equals distance divided by time; distance is twice the spring length.)

7. Calculate and record the average time and the average wave speed.

Part B—Wavelength and Frequency
Procedure

8. Keep the spring the same length that you used in Part A.

9. You will record your data in the table below.

Table 2: Wavelength and Frequency Data

Trial	Length of spring (m)	Time for 10 cycles (s)	Wave frequency (Hz)	Wavelength (m)
1				
2				
3				
Average				

10. One of the two students holding the spring should start shaking the spring from side to side until a wave pattern appears that resembles one of those shown on page 709 of your textbook.

Wave Speed, Frequency, and Wavelength, continued

11. Using the stopwatch, the third group member should measure and record how long it takes for 10 cycles of the wave pattern to occur. (One back-and-forth shake is one cycle.) Keep the pattern going so that measurements for three trials can be made.

Analyze the Results

12. Calculate the frequency (Hz) for each trial. To do this, divide the number of cycles (10) by the number of seconds that elapsed during the 10 cycles. Show your work. Record the answers in Table 2.

13. Determine the wavelength (m) using the equation given with the pattern on page 709 of your textbook that matches your wave pattern. Record your answer in Table 2.

14. Calculate and record the average time and frequency.

Draw Conclusions—Parts A and B

15. To discover the relationship among speed, wavelength, and frequency, try multiplying or dividing any two of them to see if the result equals the third. (Use the average speed, wavelength, and average frequency from your data tables.) Write the equation that shows the relationship.

Wave Speed, Frequency, and Wavelength, continued

16. Reread the definitions for *frequency* and *wavelength* in the chapter titled "The Energy of Waves." Use these definitions to explain the relationship that you discovered.

_____

_____

_____

_____

_____

_____

_____

_____

_____

_____

DATASHEET

56 STUDENT WORKSHEET

Easy Listening

DISCOVERY LAB

Pitch describes how low or how high a sound is. A sound's pitch is related to its frequency—the number of waves per second. Frequency is expressed in hertz (Hz), where 1 Hz equals one wave per second. Most humans can hear frequencies from 20 Hz to 20,000 Hz. However, not everyone detects all pitches equally well at all distances. In this activity you will collect data to see how well you and your classmates hear different frequencies at different distances.

MATERIALS
• 4 tuning forks of different frequencies • pink rubber eraser • meterstick • graph paper

Ask a Question

1. Do students in your classroom hear low-, mid-, or high-frequency sounds better?

Form a Hypothesis

2. Write a hypothesis that answers the question above. Explain your reasoning.

_____

_____

_____

Test the Hypothesis

3. Choose one member of your group to be the sound maker. The others will be the listeners.

4. Use the data table on the next page to record your data.

5. Record the frequency of one of the tuning forks in the top row of the first column of the data table.

6. The listeners should stand in front of the sound maker with their backs turned.

7. The sound maker will create a sound by striking the tip of the tuning fork gently with the eraser.

8. The listeners who hear the sound should take one step away from the sound maker. The listeners who do not hear the sound should stay where they are.

9. Repeat steps 7 and 8 until none of the listeners can hear the sound or until the listeners reach the edge of the room.

Easy Listening, continued

Data Collection Table

Frequency	Distance (m)			
	Listener 1	Listener 2	Listener 3	Average
1 (____Hz)				
2 (____Hz)				
3 (____Hz)				
4 (____Hz)				

10. Using the meterstick, the sound maker should measure the distance from his or her position to each of the listeners. All group members should record this data in their tables.

11. Repeat steps 5 through 10 with a different tuning fork.

12. Continue until all four tuning forks have been tested.

Analyze the Results

13. Calculate the average distance for each frequency, and record it in the data table. Share your group's data with the rest of the class to make a data table for the whole class.

14. Calculate the average distance for each frequency for the class.

15. Make a graph of the class results, plotting average distance (*y*-axis) versus frequency (*x*-axis).

Draw Conclusions

16. Was everyone in the class able to hear all frequencies equally? (Hint: Was the average distance for each frequency the same?)

_____

_____

_____

_____

CHAPTER 21

17. If the answer to question 16 is no, which frequency had the largest average distance?

_____

_____

Which frequency had the smallest average distance?

_____

_____

18. Based on your graph, do your results support your hypothesis? Explain your answer.

_____

_____

_____

19. Do you think your class sample is large enough to confirm your hypothesis for all humans of all ages? Explain your answer.

_____

_____

_____

_____

_____

_____

DATASHEET

57 **STUDENT WORKSHEET**

DESIGN YOUR OWN

The Speed of Sound

In the chapter titled "The Nature of Sound," you learned that the speed of sound in air is 343 m/s at 20°C (approximately room temperature). In this lab you'll design an experiment to measure the speed of sound yourself—and you'll determine if you're "up to speed"!

MATERIALS
• materials of your choice, approved by your teacher

Procedure

1. Brainstorm with your teammates to come up with a way to measure the speed of sound. Consider the following as you design your experiment:

   a. You must have a method of making a sound. Some simple examples include speaking, clapping your hands, and hitting two boards together.

   b. Remember that speed is equal to distance divided by time. You must devise methods to measure the distance that a sound travels and to measure the amount of time it takes for that sound to travel that distance.

   c. Sound travels very rapidly. A sound from across the room will reach your ears almost before you can start recording the time! You may wish to have the sound travel a long distance.

   d. Remember that sound travels in waves. Think about the interactions of sound waves. You might be able to include these interactions in your design.

2. Discuss your experimental design with your teacher, including any equipment you need. Your teacher may have questions that will help you improve your design. Describe your experiment.

   _____

   _____

   _____

   _____

   _____

   _____

The Speed of Sound, continued

Conduct an Experiment

3. Once your design is approved, carry out your experiment. Be sure to perform several trials. Record your results.

_____

_____

_____

_____

_____

Draw Conclusions

4. Was your result close to the value given in the introduction to this lab? If not, what factors may have caused you to get such a different value?

_____

_____

_____

_____

5. Why was it important for you to perform several trials in your experiment?

_____

_____

_____

Communicate Your Results

6. Compare your results with those of your classmates. Determine which experimental design provided the best results. Explain why you think this design was so successful.

_____

_____

_____

_____

DATASHEET

58 **STUDENT WORKSHEET**

Tuneful Tube

If you have seen a singer shatter a crystal glass simply by singing a note, you have seen an example of resonance. For this to happen, the note has to match the resonant frequency of the glass. A column of air within a cylinder can also resonate if the air column is the proper length for the frequency of the note. In this lab you will investigate the relationship between the length of an air column, the frequency, and wavelength during resonance.

MATERIALS
• 100 mL graduated cylinder • water • plastic tube, supplied by your teacher • metric ruler • 4 tuning forks of different frequencies • pink rubber eraser • graph paper

Procedure

1. Use the table below to record your data.

2. Fill the graduated cylinder with water.

3. Hold a plastic tube in the water so that about 3 cm is above the water.

4. Record the frequency of the first tuning fork. Gently strike the tuning fork with the eraser, and hold it so that the prongs are just above the tube, as shown on page 713 of your textbook. Slowly move the tube and fork up and down until you hear the loudest sound.

5. Measure the distance from the top of the tube to the water. Record this length in your data table.

6. Repeat steps 3–5 using the other three tuning forks.

Data Collection Table

Frequency (Hz)				
Length (cm)				

Tuneful Tube, continued

Analysis

7. Calculate the wavelength (in centimeters) of each sound wave by dividing the speed of sound in air (343 m/s at 20°C) by the frequency and multiplying by 100.

8. Make the following graphs: air column length versus frequency and wavelength versus frequency. On both graphs, plot the frequency on the *x*-axis.

9. Describe the trend between the length of the air column and the frequency of the tuning fork.

_____

_____

10. How are the pitches you heard related to the wavelengths of the sounds?

_____

_____

_____

_____

_____

DATASHEET

59 **STUDENT WORKSHEET**

The Energy of Sound

In the chapter titled "The Nature of Sound," you learned about various properties and interactions of sound. In this lab you will perform several activities that will demonstrate that the properties and interactions of sound all depend on one thing— the energy carried by sound waves.

MATERIALS
• 2 tuning forks of the same frequency and one of a different frequency
• pink rubber eraser
• small plastic cup filled with water
• rubber band
• 50 cm string

Part A—Sound Vibrations
Procedure
1. Lightly strike a tuning fork with the eraser. Slowly place the prongs of the tuning fork in the plastic cup of water. What do you see?

_____

_____

Analysis
2. How do your observations demonstrate that sound waves are carried through vibrations?

_____

_____

Part B—Resonance
Procedure
3. Strike a tuning fork with the eraser. Quickly pick up a second tuning fork in your other hand and hold it about 30 cm from the first tuning fork.

4. Place the first tuning fork against your leg to stop its vibration. Listen closely to the second tuning fork. Record your observations, including the frequencies of the two tuning forks.

_____

_____

5. Repeat steps 3 and 4, using the remaining tuning fork during the second part of step 3.

_____

_____

The Energy of Sound, continued

Analysis

6. Explain why you can hear a sound from the second tuning fork when the frequencies of the tuning forks used are the same.

_____

_____

_____

_____

_____

7. When using tuning forks of different frequencies, would you expect to hear a sound from the second tuning fork if you strike the first tuning fork harder? Explain your reasoning.

_____

_____

_____

_____

_____

_____

Part C—Interference

Procedure

8. Using the two tuning forks with the same frequency, place a rubber band tightly over the prongs near the base of one tuning fork. Strike both tuning forks at the same time against the eraser. Hold a tuning fork 3 to 5 cm from each ear. If you cannot hear any differences, move the rubber band down further on the prongs. Strike again. Record what you hear.

_____

_____

_____

The Energy of Sound, continued

Analysis

9. Did you notice the sound changing back and forth between loud and soft? A steady pattern like this is called a beat frequency. Explain this changing pattern of loudness and softness in terms of interference (both constructive and destructive).

_____

_____

_____

_____

_____

Part D—The Doppler Effect

Procedure

10. Your teacher will tie the piece of string securely to the base of one tuning fork. Your teacher will then strike the tuning fork and carefully swing the tuning fork in a circle overhead. Record your observations in your ScienceLog.

Analysis

11. Did the tuning fork make a different sound when your teacher was swinging it than when he or she was holding it? Explain why or why not.

_____

_____

_____

_____

_____

12. Is the actual pitch of the tuning fork changing? Explain.

_____

_____

_____

_____

The Energy of Sound, continued

Draw Conclusions—Parts A–D

13. Explain how your observations from each part of this lab verify that sound waves carry energy from one point to another through a vibrating medium.

_____

_____

_____

_____

_____

_____

_____

_____

_____

_____

14. Particularly loud thunder can cause the windows of your room to rattle. How is this evidence that sound waves carry energy?

_____

_____

_____

_____

_____

_____

_____

_____

_____

DATASHEET

60 **STUDENT WORKSHEET**

DISCOVERY LAB

What Color of Light Is Best for Green Plants?

Plants grow well outdoors under natural sunlight. However, some plants are grown indoors under artificial light. A wide variety of colored lights are available for helping plants grow indoors. In this experiment, you'll test several colors of light to discover which color best meets the energy needs of green plants.

▲ CHAPTER 22

MATERIALS
• masking tape
• marker
• Petri dishes and covers
• water
• paper towels
• bean seedlings
• variety of colored lights, supplied by your teacher

SCIENTIFIC METHOD

Ask a Question

1. What color of light is the best for growing green plants?

Form a Hypothesis

2. In the space below, write a hypothesis that answers the question above. Explain your reasoning.

_____

_____

_____

Test the Hypothesis

3. Use the masking tape and marker to label the side of each Petri dish with your name and the type of light you will place the dish under.

4. Place a moist paper towel in each Petri dish. Place five seedlings on top of the paper towel. Cover each dish.

5. Record your observations of the seedlings, such as length, color, and number of leaves, in your ScienceLog.

6. Place each dish under the appropriate light.

7. Observe the Petri dishes every day for at least 5 days. Record your observations in your ScienceLog.

Analyze the Results

8. Based on your results, which color of light is the best for growing green plants?

_____

Which color of light is the worst?

_____

Draw Conclusions

9. Remember that the color of an opaque object (such as a plant) is determined by the colors the object reflects. Use this information to explain your answers to question 8.

_____

_____

_____

10. Would a purple light be good for growing purple plants? Explain.

_____

_____

_____

_____

Communicate Results

11. Write a short paragraph summarizing your conclusions.

_____

_____

_____

_____

_____

_____

_____

DATASHEET

61 **STUDENT WORKSHEET**

DISCOVERY LAB

Which Color Is Hottest?

Will a navy blue hat or a white hat keep your head warmer in cool weather? Colored objects absorb energy, which can make the objects warmer. How much energy is absorbed depends on the object's color. In this experiment you will test several colors under a bright light to determine which colors absorb the most energy.

CHAPTER 22 ▲▲▲

MATERIALS
• tape
• squares of colored paper
• thermometer
• light source
• cup of room-temperature water
• paper towels
• graph paper
• colored pencils or pens

Procedure

1. Use the table below to record your data.

2. Tape a piece of colored paper around the bottom of a thermometer. Then hold the thermometer under the light source. Record the temperature every 15 seconds for 3 minutes.

3. Cool the thermometer by removing the piece of paper and placing the thermometer in the cup of room-temperature water. After 1 minute, remove the thermometer, and dry it with a paper towel.

4. Repeat steps 2 and 3 with each color, making sure to hold the thermometer at the same distance from the light source.

Temperature Data Table (in degrees Celsius)

Time:	0 sec	15 sec	30 sec	45 sec	60 sec	75 sec	90 sec	105 sec	120 sec	135 sec	150 sec	165 sec	180 sec
White													
Red													
Blue													
Black													

Analyze the Results

5. Graph temperature (*y*-axis) versus time (*x*-axis) on the grid below. Plot all data on the same grid using a different colored pencil or pen for each set of data.

Temperature Over Time for Different Colors

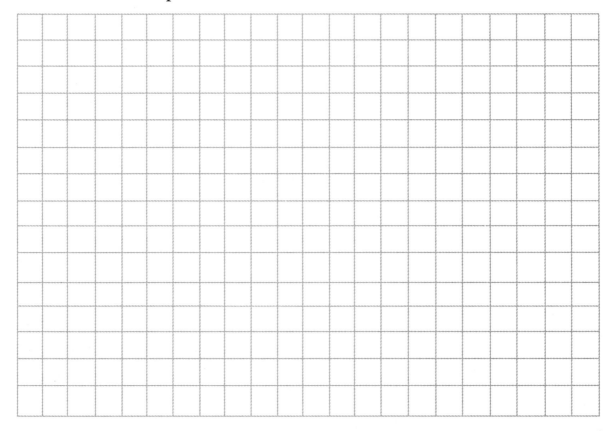

6. In the space below, list the colors you used in order from hottest to coolest.

_____

Draw Conclusions

7. Compare the colors based on the amount of energy each absorbs.

_____

_____

_____

Which Color Is Hottest? continued

8. In this experiment a white light was used. How would your results be different if you used a red light? Explain.

_____

_____

_____

9. Use the relationship between color and energy absorbed to explain why different colors of clothing are used for different seasons.

_____

_____

_____

_____

_____

▲ CHAPTER 22

DATASHEET

62 STUDENT WORKSHEET

Mixing Colors

When you mix two colors, like red and green, you create a different color. But what color do you create? And is that new color brighter or darker? The color and brightness you see depend on the light that reaches your eye, and that depends on whether you are performing color addition (combining wavelengths by mixing colors of light) or color subtraction (absorbing light by mixing colors of pigments). In this experiment, you will try both types of color formation and see the results firsthand!

MATERIALS

Part A
• 3 flashlights
• colored filters—red, green, and blue
• masking tape
• white paper

Part B
• masking tape
• 2 small plastic or paper cups
• water
• paintbrush
• watercolor paints
• white paper
• metric ruler

Part A—Color Addition

Procedure

1. Place a colored filter over each flashlight lens. Use masking tape to hold the filters in place.

2. In a darkened room, shine the red light on a sheet of clean white paper. Then shine the blue light next to the red light. You should see two circles of light, one red and one blue, next to each other.

3. Move the flashlights so that the circles overlap by about half their diameter. Examine the three areas of color. What color is formed in the mixed area?

   _____

   Is the mixed area brighter or darker than the single-color areas?

   _____

4. Repeat steps 2 and 3 with the red and green lights. Record your observations.

   _____

   _____

5. Now shine all three lights at the same point on the sheet of paper. Examine the results, and record your observations.

   _____

   _____

_____

Mixing Colors, continued
_____

Analysis

6. In general, when you mixed two colors, was the result brighter or darker than the original colors?

_____

_____

7. In step 5, you mixed all three colors. Was the resulting color brighter or darker than mixing two colors? Explain your answer in terms of color addition. (Hint: Read the definition of color addition in the introduction.)

_____

_____

_____

8. Based on your results, what do you think would happen if you mixed all the colors of light? Explain.

_____

_____

_____

Part B—Color Subtraction

Procedure

9. Place a piece of masking tape on each cup. Label one cup "Clean" and the other cup "Dirty." Fill both cups approximately half full with water.

10. Wet the paintbrush thoroughly in the "Clean" cup. Using the watercolor paints, paint a red circle on the white paper. The circle should be approximately 4 cm in diameter.

11. Clean the brush by rinsing it first in the "Dirty" cup and then in the "Clean" cup.

12. Paint a blue circle next to the red circle. Then paint half the red circle with the blue paint.

13. Examine the three areas: red, blue, and mixed. What color is the mixed area?

_____

Does it appear brighter or darker than the red and blue areas?

_____

_____

14. Clean the brush. Paint a green circle 4 cm in diameter, and then paint half the blue circle with green paint.

15. Examine the green, blue, and mixed areas. Record your observations.

_____

_____

16. Now add green paint to the mixed red-blue area, so that you have an area that is a mixture of red, green, and blue paint. Clean your brush.

17. Record your observations of this new mixed area.

_____

_____

Analysis

18. In general, when you mixed two colors, was the result brighter or darker than the original colors?

_____

_____

19. In step 16, you mixed all three colors. Was the result brighter or darker than mixing two colors?

_____

_____

Explain your answer in terms of color subtraction. (Hint: Read the definition of color subtraction in the introduction.)

_____

_____

Mixing Colors, continued

20. Based on your results, what do you think would happen if you mixed all the colors of paint? Explain.

_____

_____

_____

CHAPTER 22

▲ ▲ ▲

DATASHEET

63 **STUDENT WORKSHEET**

SKILL BUILDER

Mirror Images

When light actually passes through an image, the image is a real image. When light does not pass through the image, the image is a virtual image. Recall that plane mirrors produce only virtual images because the image appears to be behind the mirror where no light can pass through it.

In fact, all mirrors can form virtual images, but only some mirrors can form real images. In this experiment, you will explore the virtual images formed by concave and convex mirrors, and you will try to find a real image using both types of mirrors.

MATERIALS

- convex mirror
- concave mirror
- candle
- jar lid
- modeling clay
- matches
- index card

Part A—Finding Virtual Images

Make Observations

1. Hold the convex mirror at arm's length away from your face. Observe the image of your face in the mirror.

2. Slowly move the mirror toward your face, and observe what happens to the image. Record your observations below.

_____

_____

3. Move the mirror very close to your face. Record your observations below.

_____

_____

4. Slowly move the mirror away from your face, and observe what happens to the image. Record your observations.

_____

_____

5. Repeat steps 1 through 4 with the concave mirror.

_____

_____

_____

_____

_____

_____

Mirror Images, continued

Analyze Your Results

6. For each mirror, did you find a virtual image? How you can tell?

_____

_____

_____

_____

7. Describe the images you found. Were they smaller, larger, or the same size as your face?

_____

_____

_____

Were they upright or inverted?

_____

_____

_____

Draw Conclusions

8. Describe at least one use for each type of mirror. Be creative, and try to think of inventions that might use the properties of the two types of mirrors.

_____

_____

_____

_____

_____

CHAPTER 23

Mirror Images, continued

Part B—Finding A Real Image

Make Observations

9. In a darkened room, place a candle in a jar lid near one end of a table. Use modeling clay to hold the candle in place. Light the candle. **Caution:** Use extreme care around an open flame.

10. Use more modeling clay to make a base to hold the convex mirror upright. Place the mirror at the other end of the table, facing the candle.

11. Hold the index card between the candle and the mirror but slightly to one side so that you do not block the candle-light.

12. Move the card slowly from side to side and back and forth to see whether you can focus an image of the candle on it. Record your results.

_____

_____

_____

_____

_____

_____

_____

13. Repeat steps 10–12 with the concave mirror.

_____

_____

_____

_____

_____

_____

_____

Mirror Images, continued

Analyze Your Results

14. For each mirror, did you find a real image? Explain how you can tell.

_____

_____

_____

_____

_____

_____

15. Describe the real image you found. Was it smaller, larger, or the same size as the object?

_____

_____

Was it upright or inverted?

_____

_____

Draw Conclusions

16. Astronomical telescopes use large mirrors to reflect light to form a real image. Based on your results, would a concave or convex mirror be better for this instrument? Explain your answer.

_____

_____

_____

_____

_____

_____

▲ **CHAPTER 23**

DATASHEET

 STUDENT WORKSHEET

Images from Convex Lenses

A convex lens is thicker in the center than at the edges. Light rays passing through a convex lens come together at a point. Under certain conditions, a convex lens will create a real image of an object. This image will have certain characteristics, depending on the distance between the object and the lens. In this experiment you will determine the characteristics of real images created by a convex lens—the kind of lens used as a magnifying lens.

MATERIALS

- index card
- modeling clay
- candle
- jar lid
- matches
- convex lens
- meterstick

Ask a Question

1. What are the characteristics of real images created by a convex lens? How do these characteristics depend on the location of the object and the lens?

Conduct an Experiment

2. Use the table below to record your observations.

3. Use some modeling clay to make a base for the lens. Place the lens and base in the middle of the table.

4. Stand the index card upright in some modeling clay on one side of the lens.

Observations of the Images Created by a Convex Lens

Image	Orientation (upright/inverted)	Size (larger/smaller)	Image distance (cm)	Object distance (cm)
1				
2				
3				

Images from Convex Lenses, continued

5. Place the candle in the jar lid, and anchor the candle with some modeling clay. Place the candle on the table so that the lens is halfway between the candle and the card. Light the candle. **Caution:** Use extreme care around an open flame.

Collect Data

6. In a darkened room, slowly move the card and the candle away from the lens while keeping the lens exactly halfway between the card and the candle. Continue until you see a clear image of the candle flame on the card. This is image 1.

7. Measure and record the distance between the lens and the card (image distance) and between the lens and the candle (object distance).

8. Is image 1 upright or inverted? Is it larger or smaller than the candle? Record this information in the table.

9. Slide the lens toward the candle to get a new image (image 2) of the candle on the card. Leave the lens in this position.

10. Repeat steps 7 and 8 for image 2.

11. Move the lens back to the middle, and then move the lens toward the card to get a third image (image 3).

12. Repeat steps 7 and 8 for image 3.

Analyze Your Results

13. Describe the trend between image distance and image size.

_____

_____

14. What are the similarities between the real images formed by the convex lens?

_____

_____

CHAPTER 23 ▲▲▲▲

Draw Conclusions

15. The lens of your eye is a convex lens. Use the information you collected to describe the image projected on the back of your eye when you look at an object.

_____

_____

16. Convex lenses are used in film projectors. Explain why your favorite movie stars are truly "larger than life" on the screen in terms of the image distance and the object distance.

_____

_____

_____

_____

_____

Communicate Your Results

17. Write a paragraph to summarize your answers to the questions in step 1. Be sure to include the roles that image distance and object distance have in determining the characteristics of the images.

_____

_____

_____

_____

_____

_____

_____

_____